THE HATCH IS ON!

Experts Extol the World's Greatest
Hatches and the Flies They've Inspired

Chris Santella

LYONS PRESS
Guilford, Connecticut
An imprint of Globe Pequot Press

Lyons Press is an imprint of Globe Pequot Press.

Project editor: Meredith Dias
Text design and layout: Sue Murray

Library of Congress Cataloging-in-Publication Data is available on file.

ISBN 978-0-7627-8063-1

Printed in the United States of America

10 9 8 7 6 5 4 3 2 1

Contents

Contents

Acknowledgments

This book would not have been possible without the generous assistance of the expert anglers and fly tiers who shared their time and experience to help bring these great hatches to life. To these folks, I offer the most heartfelt thanks. I would especially like to thank Chris Conaty, who both offered encouragement and made many introductions on my behalf. I also wish to acknowledge the fine efforts of my agent, Stephanie Kip Rostan; my editor, Allen Jones; designer Sue Murray; and copyeditor Elissa Curcio, who helped bring the book into being.

I've had the good fortune over the last thirty years to make many fine fishing friends with whom I've had the chance to share some great hatches—and anticipate many that never quite came off! This list includes Peter Marra, Ken Matsumoto, Jeff Sang, Joe Runyon, Mark Harrison, Peter Gyerko, Tim Purvis, Geoff Roach, Kenton Quist, Mike Marcus, Nelson Mathews, Kevin Wright, John Smith, David Moscowitz, Ken Helm, Bryce Tedford, Darrell Hanks, Hamp Byerly, Mac McKeever, Robert Tomes, Conway Bowman, Kirk Deeter, and many others. I look forward to many more days on the river with these friends and friends to come.

I also extend kudos to Sloan Morris, Keith Carlson, and Doug Mateer, who've helped put fly fishing to music in our band, Catch & Release—including the title song of this book. Finally, I want to extend a special thanks to my wife, Deidre, and my daughters, Cassidy and Annabel, who've humored my absence on far too many occasions so I could chase the hatches of my dream, and to my parents, Andy and Tina Santella, who are not anglers but always encouraged me to pursue my passions.

Foreword

When you really think about it, the essence of fly fishing for trout boils down to being in the right place at the right time. We can talk about fancy casting loops and cane rods, traditions and techniques, exotic adventures and stuffed memories on our walls until we are blue in our faces. But what really, truly captures the spirit of classic trout angling is standing knee-deep in a river and watching that water in front of you come to life in a way that transcends any hope or expectation you might have had when you pulled your waders on in the first place. Let's face it: Dry-fly fishing is the top of the game. If you've seen and experienced an "epic" hatch, you know exactly what I am talking about.

I spent my growing-up fly-fishing years in Michigan, plying rivers on the west side of that state for baby browns and rainbows, and I reached a point where I honestly wondered if anything larger than the width of my hand really lived in those inland waters. But one evening, I found myself casting at dusk under the Clay Banks on the Pere Marquette River when the gray drakes suddenly swarmed and fell from the sky. At that moment, everything in my fly-fishing world changed forever. All those fish I'd only heard about . . . long, fat, wily browns that I'd assumed to be folk legends were there indeed, slurping bugs with reckless abandon from the river surface. Now, more than twenty-five years later, I still dream about that specific hatch. It's haunted me, and inspired me. I'm grateful, but I also feel cursed. Because now, I lose sleep wondering about what happens when the bugs fall from the sky on rivers farther away.

Chris Santella is the consummate angler (and writer) who has seen and experienced more than most of us might ever dream about. The premise of this book is beautifully simple: He drops us in locales at the exact times when the bugs fall and the rivers boil, as seen through the eyes of those who live to be there and experience it, from the cicada swarms described by Scott Barrus, to fishing the Skwala stoneflies with Greg Thomas. There's never been a book about being in the right place, at the right time, quite like this one. It is, indeed, the substance that fuels the most intense, and longest-lasting, fly-fishing dreams.

—Kirk Deeter
Editor-at-Large, *Field & Stream*
Editor, *Trout*

Introduction

It's always good to get out on the water to fly fish. But there are certain times when anglers can expect a very special experience—times when bugs are in the air, and trout throw all caution to the wind to chow down.

Times when *The Hatch Is On!*

These are the times anglers dream of through long, cold winters, times that anglers mark on the calendar in bright highlighter pen and try to plan annual trips around. They speak in hushed voices around the fireplace or over a pint at the local pub of the Skwala stone in April, the green drake and salmon fly in June, the blanket caddis and Hexagenia in July, and other epic hatches—moments when the fishing can be so good, it creates memories of a lifetime. Even the hatches that don't come off create good times, as they've provided an excuse for friends to gather and partake of a shared passion.

The Hatch Is On celebrates some of fly fishing's most treasured insect emergences (and one mammal occurrence) and the incredible angling experiences these hatches create. To compile the book, I interviewed accomplished guides and fly tiers from around the country and asked them to discuss a favorite or memorable hatch, why it's exciting to them, and how

Blanket hatches are always impressive. But for those anglers lucky enough to catch a massive emergence of caddis flies, few dry-fly experiences can compare.

it's informed their activities at the bench. *The Hatch Is On* is not intended to be the final word on BWOs or PMDs; instead, each essay reflects the experiences of the anglers in question. It's my hope that the book will provide some insights into how talented tiers approach their craft, and some ideas for planning your next trout-fishing escape.

I hope you can catch a few of these great hatches in the years to come!

Ants thrive around dead trees, like those found at Crane Prairie Reservoir in central Oregon. BRIAN O'KEEFE

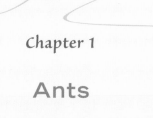

Chapter 1

Ants

Ralph Cutter

Ants are the world's most numerous insect. It's been estimated that for every human walking the earth, there are 7,000 ants. More than 1,000 species can be found in the United States. Despite their prolificacy, they may not be the first bug trout anglers think of tying on as they set out on the water. Sometimes they're even overlooked by the most astute observers of insect life.

"My wife Lisa and I were out in a pram on Milton Reservoir [north of Truckee, California]," Ralph Cutter recalled. "We'd been fishing for three hours, and we hadn't found a single fish. We'd tried nymphs, emergers, dries. Nothing. Meanwhile, a guy in a float tube about 100 feet away was playing a fish every five minutes. He seemed to be using something on top. I couldn't bring myself to ask what he was using. Finally, Lisa asked. He said, 'I'm using a Cutter's Ant.' I had been keyed in to what was *supposed* to be working, from a mayfly perspective. I'd momentarily forgotten all the dying, rotting lodgepole pines in and around the water—perfect ant habitat. It was a case where we needed to *unmatch* the hatch.

"I think ants are the killer of all terrestrials. Being as abundant as they are, it's only natural that they're the insect that's most frequently encountered by trout. I see them as a universal pattern, a go-to fly. They look like beetles and snails as well as ants. You can use them anywhere, though my favorite places to fish ants are lakes with dead and dying trees—whether the decaying trees are a result of beavers or Corps of Engineers–manufactured impoundments. Carpenter ants live in those trees, and they're falling into the water year-round.

"This being said, there are periods when you'll see great swarms of flying ants when the bugs, carried on temporary wings, set out to find suitable habitat to establish new colonies. Living in Truckee, there are times—usually in early summer, during the first really warm days—when you can live by the ant hatch. You can hear them hitting against the windows of the house and see them splattering against the windshield. It gives anglers in the know a primal, Pavlovian response! My experience has been that the fishing is typically not great when the hatch starts. The ants are big, a size 8 or 10. At first the fish seem wary, almost afraid of them. But once they figure things out, the trout are on them, and they'll continue eating artificials for a week or two after the emergence is done."

Ralph is the first to admit that the plethora of foam ant patterns on the market can work quite well. But these imitations float high on the water, and Ralph prefers his ants to rest a bit lower in the water. "I had an 'ah-ha' moment a few years back on Hot Creek [on the east side of the Sierras, near Mammoth Lake]. Carpenter ants were dropping on the water in large numbers, but the trout weren't eating them as they floated past. I knew the fish were in there, and once I got a fly down a bit, the fish started killing it. The trout were waiting for the ants to drown. This highlights a common mistake many people make with a number of dry flies—they want bugs on top where they can see them better. The fish want them further down."

When Ralph set out to create what he hoped would be a better ant pattern, he began by looking at his subject from underwater. "As you view ants from below, you notice that they're covered with little hairs," he continued. "The hairs, or bristles, tend to trap little bubbles of air. These give the ant a silvery brightness from below, even though they're black on top. Lots of the patterns I've seen use regular dubbing and lack this brightness, this sparkle—though that's not always true of the foam ants, some of which have little pockets that capture air and glisten. I started experimenting with Antron, which traps air and has some sparkle. Antron doesn't float very well, however, so I tied in some deer hair on the top half of the pattern Humpy style, a shell of sorts. The hollow-celled interior of the deer hair helps float the ant pattern in the film.

"The tips of the deer hair are posted upright, and I wrap a few turns of hackle around them, parachute style. This hackle helps hold the ant in the film and dents the water identically to the dent of the legs [and wings, if present] of the natural. It takes me about a minute to tie, it has few parts, and it works. I named it the Perfect Ant—not because it's even close to perfect, but because they say that practice makes perfect, and I certainly had lots of trial and error to come up with the pattern."

If there's a lake that was custom-made for fishing ants, it's Crane Prairie Reservoir, southwest of Bend, Oregon, in the shadow of Mount Bachelor. Crane Prairie was created in 1922 with the damming of the upper Deschutes River; hundreds of thousands of lodgepole

Flying Ant BRIAN O'KEEFE

Cutter's Perfect Ant BRIAN O'KEEFE

pines were left standing as the waters flooded this 5-square-mile meadow. Many have sunk to the bottom, but a fair amount still stand. The shallow lake is home to some outsize rainbows, lovingly dubbed "Cranebows" by locals. These hard-fighting fish can push 10 pounds; fish approaching the 20-pound mark are seen most years. "I was fishing dragonfly nymphs [an important food source on the lake] near some lily pads on Crane Prairie," Ralph recalled. "In a cove nearby, the surface of the water started moving. It was obvious a big hatch of something was coming off. I pushed over and saw swarms of ants spewing out of nearby stumps like a flume of smoke. Many were falling in the water, and the fish were beside themselves."

Cutter's Perfect Ant

Hook: TMC 9300, #12–16

Thread: Black 6/0 Danville

Body: Black or rust Antron

Legs: Brown hackle

Shell: Black or red deer hair

Thorax: Black or rust Antron

Ralph Cutter operates the California School of Fly Fishing (www.flyline.com) with his wife, Lisa, from Nevada City and Truckee, California. He is the author of the award-winning *Sierra Trout Guide* and *Fish Food*; he and Lisa have produced the DVD *Bugs of the Underworld*.

Most fishing on California's Fall River is done from small prams. Downstream presentations are the order of the day, as fish are extremely leader-shy. JOHN SHERMAN

Chapter 2

Blue-Winged Olives

Bob Quigley

"When I see a particular kind of cloudy weather in the spring or fall—and sometimes even in the summer or winter—I know it's time to get to one of my favorite blue-winged olive streams," Bob Quigley declared. "It might be the Fall River down in northern California or the Holy Water on the Rogue, not far from where I live in Oregon. The hatches can be so prolific, you feel like you should be getting them on every cast. But instead you might get only one or two. There are some days when they kick your butt."

Blue-winged olives may be the world's most misunderstood mayfly—literally! Fisheries biologist and TroutNut.com founder Jason Neuswanger has noted that "*Baetis* is probably the most misidentified genus in the angler's mayfly world. Many fly anglers see anything too small to imitate with a size 16 Adams and call it *Baetis*. In reality, *Baetis* is the most prominent of several very similar abundant genera in the family Baetidae. It seems every species in the family is perpetually being reclassified, and identifying any of them, even to genus level, is difficult."

If the angling family can't agree on exactly what a BWO is, we do know that these small mayflies are extremely prolific on North American rivers, tend to show up in cooler months (especially in the West) on cloudy days, and are frequently neither green-bodied or blue-winged. In lieu of misnaming the bugs, Neuswanger suggests simply calling them "little [whatever color they are] mayflies."

Why do we associate BWOs with cloudy/rainy days? There's been no shortage of speculation on this point. Some have posited that the duns emerge during lower light so their wings do not dry out. Proponents of this theory go on to point out that BWO duns tend to have darker wings in the winter/early spring, which would help absorb heat during cooler times, and lighter wings as warmer weather approaches, which would deflect solar heat. Others have proposed that BWO hatches vary little between sunny and cloudy days when the bugs are present, but that fish are much less active on bright days, perhaps fearing easy exposure to winged predators.

"My earliest memories of fishing BWOs are when I was 13 or 14," Bob continued. "I spent a lot of time with relatives in Idaho and Montana, and I recall fishing on Silver Creek, Henry's Fork, and the Madison. I remember going into a fly shop in Sun Valley and getting some Adams-type patterns, with standard hackle. I learned the proper presentation methods, and they worked. The adrenaline rush of my first big fish on a Baetis—a 24-inch brown taken from a weedy channel on Silver Creek—was profound. I've done a lot of fishing since then, have guided and started a fly shop. Now I study the insects closer." This level of observation helped bring the Quigley Cripple to light, a style of pattern that mimics mayflies struggling to break free from their nymphal shuck—and has been mimicked by many tiers since.

Another Bob Quigley innovation that's come to bear on many mayfly patterns—including BWOs—is the Stacker style. Fishing on the clear, smooth spring creek waters of the Fall River, Bob wanted a fly that would land lightly on the water, have a true mayfly silhouette, and remain visible to the angler. To achieve this, he developed a revolutionary way to apply hackle to the fly. Chris Conaty, director of product development for Idylwilde flies, described Bob's technique this way: "Instead of palmering hackle around the shank of the hook or a parachute to create floatability, Bob parachutes the entire hackle up a monofilament stacker loop and then pulls the entire loop over the thorax. This encases the top half of the thorax, and creates a stacked dome of hackle barbules on top of the fly."

The result is a "twofer," as the stacked hackle imitates the mayfly wing silhouette of both an upright or spent spinner and an upright or fluttering dun. Bob testified to the Stacker's efficiency: "I had an experience once on the Fall River one cloudy morning in

June. On the left side of the river, pale morning dun and BWO spinners were on the water. On the right side, there were emergers and duns. It was pretty satisfying to catch fish on both sides with the same fly."

Bob offered another tip: "You just can't get your wings dark enough for a dark, cloudy day. You almost want to go to black. If you take a close look at a dun, you'll notice that there's more brown in their bodies than you'd realize."

Bob described a typical day during the BWO emergence on the "Holy Water," a small but extremely fecund section of the Rogue River below Lost Creek Dam that's managed for fly fishing. "I like to get there around 10. Nothing will happen until 11:30 or 12, but I scout around a bit and get my bugs together. Some guys will fish indicators before the hatch begins, but I like to wait until the fish are up. The Holy Water has high banks, and if you know the bug lanes from experience, you can hone in and look for swirls. Once I spot a swirl, I'll check my fly and leader one more time and wait until some of the bigger fish I've spotted get in a rhythm; I want them to be happy when I put the bug out. I'll slowly slide into the river and get some line off the reel. I'm using a long leader, 14 to 18 feet, and a light rod, as the heavier line weights will splash. Over the years, I've learned to tie leaders that work almost like line. Tippet will usually be 6X or 7X, and

I'll fish a size 20 or 22 pattern, maybe a Hackle Stacker or a Stacker Flag Dun. Some guys fish flies to size 26 and 28, but 22 is generally the break-off point for me.

"I try to make as short of a drift as I can, to land the fly in the fish's ring. I think you spook fish with longer drifts. If I don't get a take with the first few drifts, I'll wait a little while. When I start in again, I might change flies or I might not. I've been at this long enough that I know when things are right—when the fish are feeding intensely and I'm getting the right drifts. If I don't get a look when things are right, I haven't got the right pattern.

"Some days you can't zero it in as much as you'd like, but some days you can. It depends on how gluttonous the fish are. On those good days, it's rewarding

Blue-Winged Olive JEREMY ALLEN

BWO Flag Dun BRIAN O'KEEFE

to find a half a dozen rainbows that are near or over 20". It's even better to be able to pass the fly that was working on to someone else you meet on the river."

BWO Flag Dun

Hook: TSF 100, #20 or #22

Thread: Camel brown or dark olive 8/0 or smaller UNI-Thread

Tail: Dark dun micro fibets

Body: Camel brown or dark olive

Hackle: Dark blue dun

Wing: Black poly or dark dun

Thorax: BWO dubbing

To make this pattern a Sparkle Flag Dun, use dark brown Z-lon for the tail instead of fibets.

Bob Quigley was one of the most influential and innovative fly tiers of the last few generations. Among his many "out of the box" contributions are the cripple, paranymph, loopwing parachute, hackle stacker, and egg-sucking concepts.

Artist, Scientist ... Fly Tier

Bob Quigley passed away from cancer on June 12, 2012, shortly after being interviewed for this book.

Bob was taken with fly fishing at an early age. After attending Humboldt State University in Arcata, California, Bob headed a few miles east to Fall River Mills. The Fall proved a wonderful observation area and testing ground for his unorthodox (but highly effective) creations. Bob was ever generous in sharing his knowledge and expertise, and no doubt inspired many anglers to try their hand at tying.

A few of Bob's contemporaries shared their thoughts on his great contributions to fly tying:

> "Bob was surely one of the most original fly tiers I have ever seen, heard, or read about. His innovations will stand the test of time. They already have. Whenever anyone thinks of spring creek fishing they should tip their hat to Bob and his flies. His hackle stacking technique and crippled mayfly designs were and are so crazy good that they are hard to improve upon."
>
> —*Chris Conaty*
> *Director of Product Development, Idylwilde Flies*

> "What separates Quigley in my mind from the masses of exceptional tiers is his ability to make major if not paradigmatic shifts in fly design. No one that I know contests the fact that Bob created the first mayfly cripple in 1978. Never before had the concept of a half dun, half nymph fly been so deliberately and effectively designed."
>
> —*Ken Morrish*
> *Celebrated fly tier, co-founder of Fly Water Travel*

Brown drakes come off in the East and West, including significant emergences on Idaho's Henry's Fork.
JOHN JURACEK

Chapter 3

Brown Drakes

Charles Meck

"The first brown drake hatch I ever saw was on Pine Creek [in north-central Pennsylvania] in 1973," Charlie Meck recalled. "It was the heaviest hatch—the greatest concentration of insects—I'd ever witnessed. I was fishing with a friend, and we started seeing a few of the bugs on a tributary to Pine Creek. Another angler we spoke to on the river thought there was going to be a green drake emergence. At the time, no one had identified the brown drake as a hatch of any importance in the Northeast. Pretty soon, there were so many spinners in the air that you could hear the loud humming of their wings. There were a minimum of 50 trout, as many as 150, feeding . . . and we didn't have a match for it! I tied on cream-colored March Browns, Black Quills. We kept switching flies, but nothing worked. The fish were very selective, to say the least. After an hour, each of us had caught one fish. This was one of the few times I can remember when I was actually glad that the hatch was over; that was how frustrated I was. They say you forget about bad experiences in your life, but this one has lingered on. One of my last impressions on leaving the river that evening was a blooming purple rhododendron.

"A few years later I happened upon brown drakes on the Henry's Fork in Idaho, another immense hatch.

I recall coming back to the car in near darkness, passing through a park ranger's front yard. In the fading light I could make out a purple rhododendron. It's the flora I always associate with the hatch."

The brown drake—*Ephemera simulans*—is a larger mayfly (most tie imitations on #10 or #12 hooks) that appears on a variety of rivers from coast to coast. In some parts of the United States, it can be the first big bug of the year to appear, and as such, can put normally wary trout off-guard. (For Midwestern anglers, it's a precursor of the big mother of all mayflies, the Hexagenia.) "I've seen brown drakes come off in big tailwaters like the Delaware and Henry's Fork, spring creeks like Silver Creek in Idaho, and little streams like Honey Creek in south-central Pennsylvania, which is only 15 feet wide in some places," Charlie continued.

Like the brief lives of mayflies, the brown drake hatch is ephemeral. "It doesn't last for very long," Charlie added. "You might hit it four or five evenings in a row on many systems, and then it's gone. The first hatches occur on eastern streams between May 25 and May 30; the bugs come off around June 10 in the Midwest, around June 25 in the western states, and the Fourth of July up in Labrador. It's pretty reliable; you can plan a trip around it." And many do. If the planets align correctly, the brown drake can make for some very memorable dry-fly fishing. While fish will take duns that emerge with considerable force from the film, it's the frequently massive spinner fall that fish really target.

The brown drake emergence is generally an early evening event. "As daylight fades, male spinners begin their mating flight, usually near trees," Charlie described. "Once females begin to congregate above the surface, the males descend. Not long after, the females lay helpless on the surface. You're not quite fishing in the dark. You can still make out the splashes of fish feeding." Silver Creek, near Sun Valley in central Idaho, is one river where the bulk of the action unfolds under the cover of darkness. It's one of the times when the creek's biggest browns, which can eclipse 2 feet in length, leave the safety of cut banks and logjams to feed on top. The legions of head-lamped anglers making their way along the river at the Nature Conservancy preserve can resemble a post-midnight climbing party attacking Mount Kilimanjaro!

In Charlie's experience, the key to successfully replicating brown drake naturals rests in matching the bug's color scheme. "The body color on the dun and spinner is elusive—a tannish brown, with a suggestion of olive sheen," he explained. "To take the guesswork out of the process, I've come up with two new concept patterns: the Color Matcher and the Quick Trim. The Color Matcher is tied with all white materials—white wing, white body, white hackle. I carry them in a variety

Brown Drake NICK PRICE

Brown Drake BRIAN O'KEEFE

of sizes—8 through 16—though for the brown drake on my home waters in Pennsylvania, a size 12 with a long-shank hook is about right. I also carry an assortment of nontoxic permanent markers, the kind you can find at an office supply store. The idea is that I can use the markers to mimic the bugs in real time. I particularly like the Bic brand. Their brown and tan are perfect for capturing the look of the brown drake.

"The Quick Trim is a parachute-style dry that I tie with extended bodies using either vernille or white closed-cell foam. The foam is the kind used in packing material; it will float for ten years even weighed down with lead shot! The bodies and the parachute post wings can be trimmed on the river to match the size and profile of the naturals. Again, the foam patterns can be colored with markers to match the bugs.

With the Color Matcher and the Quick Trim, a set of markers, and some scissors, I'm never without a match for what's happening on the river."

Quick Trim™ Dry Fly
Hook: Standard dry fly, #14 or #16
Thread: White 6/0
Wings: White poly yarn
Body: White vernille (comes in 2 sizes; use the larger one for #14 and bigger, and the smaller one for #16 and smaller), thin-celled packing foam, or white poly foam cut in thin strips. Make body and wings long enough to copy very large flies (#6).
Hackle: White saddle hackle, tied parachute, colored with permanent marker (Bic comes in dozens of colors)

Make sure to bring a selection of permanent markers with you. If all of a sudden a yellow drake appears, color the fly with a yellow permanent marker and cut off the back end of the body and wings to size.

Charles R. Meck's numerous books include *Trout Streams and Hatches of Pennsylvania* (third edition), *Arizona Trout Streams and Their Hatches*, *Mid-Atlantic Trout Streams and Their Hatches*, *Fishing Small Streams with a Fly Rod*, *Meeting and Fishing the Hatches*, and *Matching Hatches Made Easy*. He lives in Pennsylvania Furnace, Pennsylvania, and Mesa, Arizona.

Caddis are perhaps the most dependable hatch on Oregon's
Deschutes River in the late spring and summer. BRIAN O'KEEFE

Chapter 4

Caddis

Brian Silvey

"Caddis are the cheeseburger, the protein shake, of many western waters," Brian Silvey began. "It's certainly true on my home river, the Deschutes. There will be times when there are no bugs on the water, but the rainbows will still take imitations. It's not a coincidence; in late spring and summer, the fish always have caddis in some form in front of them."

Caddisflies (sometimes referred to as sedge flies as one moves east) are a significant insect order (Tricoptera), with some 1,200 species in North America alone. Their casings—little homes the bugs construct from rocks, twigs, and other debris—are ubiquitous on many streambeds. (Inspired by the insects' work, a stream ecologist named Ben Stout and his wife, Kathy, have even created a line of caddis casing jewelry!) Some of us have no doubt encountered a "blanket caddis hatch" where it's difficult to draw a breath without taking in several bugs. Sometimes the bugs will settle on the water and an Elkhair Caddis in tan (or brown, or green, or *your color here*) will knock 'em dead . . . if you can make out your fly in the dying light when the egg-laying/spinner fall occurs. But frequently those blankets will prove to be food only for the angler, as the caddis swarm merely moves upstream.

Considering their abundance and geographic scope, caddis seem deserving of the kind of attention that's lavished on the more gracefully streamlined mayfly. Yet before the late Gary LaFontaine came along, caddis were largely misunderstood, and perhaps for this reason, given shorter shrift. ("It doesn't help that many people mistake them for moths," Brian quipped.) While attending the University of Montana at Missoula, LaFontaine donned scuba gear and spent countless hours underwater observing caddisfly behavior as he worked toward a degree in behavioral psychology. Perhaps more significantly for our purposes, he carefully noted how trout behaved when presented with an offering of caddis. His notes and interpretations were immortalized in *Caddisflies,* the 1981 title that's still the definitive book on the topic.

One of LaFontaine's novel observations was that during the early stages of a caddis emergence, trout chasing pupae would often break the surface of the water. Not a few anglers have mistaken this behavior as surface feeding, and have gone through every fly in their box trying to decipher what bug the fish are taking. LaFontaine explained that this surface activity was their momentum in chasing the pupae that carried them out of the water, or their backs/tails bulging as they took bugs in the film and headed back down. (Takeaway note: If it's summer and you see fish rising but no bugs are evident, a caddis emergence is under way!)

Another observation was that when many species of caddis pupae rise to the surface, their abdominal areas are surrounded by an air sac—and that the trout seemed to be attracted to the bubbles. One of the most storied caddis imitations, the LaFontaine Sparkle Pupa, was created to capture this moment of emergence. A key element in the design of the fly was a nylon fiber developed by DuPont called Antron, which was notable (per DuPont) for its "tighter molecular structure, which gives the fiber unsurpassed strength and utility in high traffic areas." Where DuPont would go on to incorporate Antron into a variety of carpets, Gary LaFontaine used it to form the pupa case in his emerger. He once explained DuPont's collaborative role in the process while conducting a seminar for the Nat Greene Fly Fishers in Greensboro, North Carolina: "DuPont was incredible. They had four to five of their top engineers talking to me. When they told me they only sold Antron by the carload, I said I'd drive down and they said 'We mean by the *train* carload.' But they sent me a lot of it."

"I used to fish LaFontaine Sparkle Pupas a lot on the Deschutes, as the fish really key in on caddis emergers," Brian shared. "But I could never tie it right.

I wanted to find a pattern I could tie fast—'guide flies,' I like to call them—so I can spend less time tying and more time fishing. I began playing with some different materials and came up with the Edible Emerger. The body is slim and tapered at the end, and I added marabou to add movement to the pattern, a material none of the existing patterns use. Fish key on movement, and marabou has very lifelike qualities when it's wet. The deer hair head is there to keep the pattern fishing on the surface or just under the surface. It has the added effect of creating a bubble trail when swung on a tight line.

"Something I really like about the fly is that you can fish it in a number of ways. You can grease the deer hair head so it stays above water and the rest of the fly is in the film or just under the surface. Or, you can tie it off the back of a high-floating caddis dry like an Elkhair Caddis. If you fish it this way, be sure to swing the fly to shore at the end of the drift, as a lot of takes come on the swing. You can also nymph it, but make sure it's the upper nymph on a two-fly rig so that it floats higher in the water column. If you see a fish rise while fishing this way, cut off your second nymph and take any weight off the rig. Now you can cast the Edible Emerger at the rise without having to completely re-rig."

As any serious tier can attest, there are times at the vise when you know where you want to go and you're trying to find better ways to get there. And there

Caddis JAMES ANDERSON

are other times when the fish are trying to tell you what to do, and your job is to interpret their bidding. Brian described one such caddis conundrum on the Deschutes: "I had been guiding for a number of days on the lower river, on a stretch of river above Mack's Canyon. There was a spot we'd reach on the float mid-afternoon—around 2:30 or 3—where there were great numbers of spent tan caddis on the surface. The trout were rising all over, but we couldn't catch any. I tried some Goddard Caddis and other spent caddis patterns. Nothing. I tied on some Elkhair Caddis and cut them way down, and we found a fish or two, but nothing to speak of.

"That night I sat down at the vise and came up with a slightly different approach. On the next day's float, the spent bugs were on the water at the same spot and the fish were on them, but I had similarly poor results. The challenge, it seemed, was the wings. Most low-riding spent caddis have a hackle or hair wing. These ties don't give the same silhouette as what we were seeing on the Deschutes. I started experimenting with some plastic (Aire-Foam) wings, and results started improving. By the third or fourth day, we had the right pattern—Silvey's Dead Caddis. The fish were almost fighting to get to the fly. In the course of a week, we went from no fish to a few fish to hooking almost every fish in the run."

Silvey's Dead Caddis KEITH CARLSON

Silvey's Dead Caddis

Hook: TMC 900 BL, #14

Thread: Brown 8/0

Antennae: 2 brown micro fibets

Wing: Tan Spirit River Wings & Things

Dubbing: Tan Super Fine Dry Fly Dubbing

Hackle: Brown

Brian Silvey grew up on Oregon's Deschutes and Sandy Rivers, and now operates a year-round guide service, Silvey's Fly Fishing (www.silveysflyfishing .com). A private and commercial fly tier for almost thirty years, Brian's patterns have been featured in books and magazines, and many are now produced by Idylwilde Flies.

If you fish still waters in the West, you know Callibaetis, which fool fish from the southern Rockies to the Kamloops region of British Columbia (shown here). BRIAN CHAN

Brian Chan

"I grew up fishing for salmon with my dad off the coast of Vancouver, British Columbia, where I grew up," Brian Chan began. "One day, a neighbor who had a son my age asked me if I wanted to go trout fishing in the interior. I went along, and we fished with bait . . . and my infatuation with lake fishing began. The infatuation grew into a passion, and I eventually decided I wanted to work around the fishing world. When I came out of college with my degree in fisheries management, the province had several fisheries management positions available—including a spot in the Kamloops region. That was thirty-seven years ago, and I've been here ever since."

For someone who enjoys fly fishing for trout in lakes, it doesn't get much better. Kamloops is world-renowned for its small-lake trout fishery and its "Kamloops rainbows." "Most of the lakes in the interior regions of British Columbia are shallow, nutrient-rich basins," Brian continued. "These lakes support abundant and diverse insect populations. Thanks to good management policies—including the introduction of triploid rainbows [sterile fish that grow at above-average rates] the lakes have good numbers of trophy fish. There are a hundred lakes I can fish within one hour of my house."

And Callibaetis come into play on all of them.

If you fish on lakes anywhere from the Rocky Mountains on west, you know Callibaetis . . . or you should. "They are the most abundant mayfly species in western still waters," Brian continued. "And on the lakes around Kamloops, they provide the first dry-fly fishing of the year. Depending on the water temperature of the lake, it can begin anytime from mid-May to the end of June. It's not a long emergence; we'll see the majority of the hatch over ten to fourteen days. But once it starts it's very reliable, very intense, and will come off on successive days. It's a very gentlemanly hatch, too—they usually begin coming off around 11:30, so you can sleep in."

Brian described a typical day when the Callibaetis are helping anglers from British Columbia and beyond shake off the doldrums of winter and early spring. "I'll usually begin fishing some shoals, never deeper than 15 feet. I like to have two rods rigged—one with floating line and a very long leader, the other with floating line and a shorter leader. I'll start with the longer leader setup and tie on a gold beadhead Hare's Ear. It's pretty generic, but the pattern pulsates nicely and looks close enough to what the fish were eating the day before. Assuming I catch a few, I'll do a quick throat pump. When I see nymphs showing up in the fish's throat, I try to get more imitative. I like a Pheasant Tail nymph tied with Hungarian partridge legs. I'll start with beadhead Pheasant Tails, but as the fish start to get filled up, they won't eat a beadhead. As the hatch progresses, your imitations need to get more and more slender, which is a more realistic imitation. You also have to be conscious of the color of the naturals. Since most of our lakes are clear, the Callibaetis nymphs tend to have dark backs and a light tan, almost white underbody. If you don't keep your ties slender and capture the two-toned color scheme, you're not going to do as well.

"If all goes right, you'll start to see some fish porpoising as they key in on emerging Callibaetis and adults. Now things can get exciting. Days with overcast skies—and even better, a little drizzle—are best. The fish feel safer coming up to take the duns. On a bright sunny day, they won't show themselves. At this point, I'll exchange the rod with the long leader for the shorter leader setup, so I can lay down a softer cast. I'll probably have a Parachute Adams tied on. The Adams is a perfect double-duty fly for the Callibaetis emergence: If the fish are eating emergers, I'll apply a little floatant just to the post of the fly so the body will sit low in the surface film; if the fish are on duns, I'll tie on another Parachute Adams and apply floatant to the body so the fly rides a little higher. A grayish body seems to work best.

"There are two ways I like to present the fly when the fish are up, depending on how they're behaving. If there are lots of fish concentrated around the boat, I like to track individual fish. After it takes a dun, you can guess that it will advance 3 or 5 feet. I'll try to drop the fly there, pull the slack line out, and retrieve the fly a few feet and let it sit . . . and wait for the bulge. If the fish are more scattered and there's a breeze, I like to wind drift. You do this by

Callibaetis JOHN JURACEK

quartering a cast perpendicular to the wind so your line, greased leader, and fly get pushed downwind. If you let a small bow develop in the line, you'll get an almost drag-free drift. The fly will ride the waves, and you'll cover lots of water."

Many of the lakes in the Kamloops region boast crystal clarity, and one day this clarity provided a rare and mildly humbling window into just how particular local rainbows can be about their Callibaetis imitations. "It was a sunny, calm day, and I was out on Lac Des Roche with a friend," Brian recalled. "The lake has numerous shallow shoal areas, and I was fishing a section that has a light-colored marl bottom. It resembles a bonefish flat—except this flat is

Stillwater Callibaetis BRIAN CHAN

8 to 17 feet deep. Looking down, I could see legions of trout. There were several groups of fish in the 3-pound class, with three to five fish in each group. There were also a number of much larger fish in the 7- to 10-pound class circulating by themselves. The water at this spot was about 8 feet deep. We cast out Pheasant Tail nymphs and watched the whole scene unfold. The groups of smaller fish would slowly track the fly, very wary and hesitant, until one would dart out and grab it. We saw many of the bigger fish inspect the fly at length, but they wouldn't take it. We landed some lovely fish that afternoon, but the big fish were beyond us. Had there been a little cloud cover or a little wind to put a ripple on the water, we might have had a chance."

Stillwater Callibaetis Nymph

Hook: Nymph, #14/2X or #16/2X (e.g., Mustad R72)
Thread: Olive 8/0 UNI-Thread
Tail: Natural Hungarian partridge breast-feather fibers
Body: Light olive ostrich herl
Shellback: Hen-pheasant tail fibers from center tail feathers
Rib: Extra-fine gold wire
Throat: Natural Hungarian partridge breast-feather fibers

Brian Chan is the author of *Morris and Chan on Fly Fishing Trout Lakes* (with Skip Morris), *Stillwater Solutions Recipes—30 Proven Patterns* (with Phil Rowley), and the DVD *Expert Techniques for Stillwater Fly Fishing*. He is a co-host of *The New Fly Fisher Show* and also guides (www.riseformflyfishing.com) in the Kamloops region.

A Kamloops Trout—Or Just a Rainbow?

The lakes of the Kamloops region of southern British Columbia have attracted generations of anglers, both for their fine stillwater fishing as well as the chance to tangle with the famed Kamloops trout. First encountered by Anglos in the larger lakes around Fort Klamath, Kamloops trout were renowned for both their size and fighting ability. Even into the 1930s, fish of fifteen pounds and more were not uncommon in B.C. Some have even questioned whether Kamloops are rainbows or a separate species.

Initial studies conducted in the 1890s by Dr. David Starr Jordan (at Stanford University) seemed to confirm that Kamloops trout were indeed a new species. Several characteristics set them apart from common specimens of *Oncorhynchus mykiss* (then classified as *Salmo gairdneri*), including:

- The fish averaged 150 to 154 rows of scales, many more than rainbows.
- They had fewer gill rakers.
- They had fewer bones in their dorsal and anal fins.
- Their heads were wider and longer.

For a time, *Salmo kamloops* entered the lexicon as a legitimate species. But in the early 1930s, a Biological Board of Canada biologist named Charles Mottley began conducting studies on the Kamloops trout. His analysis showed that the variations between *Salmo kamloops* and other rainbows were due to environmental conditions rather than genetic differences—specifically, the fact that the spawning streams in south-central British Columbia were approximately nine degrees cooler than typical rainbow spawning habitat.

The canyons of the Green River below Flaming Gorge Reservoir provide a dramatic backdrop for the almost annual cicada emergence. BRIAN O'KEEFE

Chapter 6

Cicadas

Scott Barrus

Cicadas are the terrestrial that are often heard but seldom seen, as least by humans. Their clicking or buzzing noise (created by their tymbal, a vibrating membrane) is a harbinger of summer's arrival in many locales. At lengths up to 2⅜ inches, they pack a load of protein. While not encountered as frequently on the menus of discerning trout as grasshoppers, cicadas occur in pockets around the United States, including emergences in Michigan and Arkansas. Their most significant occurrence comes each spring on the Green River in northeastern Utah. Come June, cicadas emerge from underground burrows, climb the trunk of a nearby tree, and shuck their skins. Their membranous music is indeed a siren song to trout and anglers alike around the small fishing outpost of Dutch John.

Cicadas are the largest member of the suborder Homoptera, which include leafhoppers, treehoppers, and aphids. Females lay their eggs in slits made in twigs of trees and shrubs. After emerging, nymphs fall to the ground at the base of the tree, where they burrow under the soil. They live underground and feed on the sap of the roots of trees and shrubs. Periodical cicadas (genus *Magicicada*)—sometimes known as 17-year cicadas east of the Mississippi—grow very slowly, taking anywhere from thirteen to seventeen years to reach adulthood. These cicadas have one of the longest life spans of any insect, if you can call their long subterranean sabbatical *living*. Grand western cicadas—the species encountered in Utah—have a shorter maturation period, though specifics are still something of a mystery. "The life cycle of western species is poorly understood," according to Alan Roe, an insect diagnostician at Utah State University. "It's believed that western species probably have a life cycle of two to five years. Regardless of this, emergence of some adult cicadas occurs annually in many types, such as the genera *Platypedia,* which is found at lower elevations in pinyon-juniper communities or shrub-land communities in association with oak or mountain mahogany."

"I began working on the Green River when I was 19," Scott Barrus recalled. "That first year I ran shuttles, but I got on the river a fair bit. The cicada hatch was on when I arrived from college, and I was blown away. I'd never seen anything like this—such big bugs on the water, and the fish going crazy. It was awesome to cast something that big, have it go 'splat' on the water, and have the fish on it."

Emmett Heath, regarded as one of the Green's most knowledgeable guides (dubbed "the Dean of the Green" by admirers), began fishing cicadas in the early

'80s. "The first one I ever tied was right on the river. The bugs were heavy and I happened to have my kit with me, so I tied up a bug with a spun black deer hair body and an elk wing. A guy I was with was catching fish right and left on the fly. Another fellow on the river was wondering what we were using. I tied him one, and we fished 'em from that day on." Emmett and his team experimented using black braided rope for bodies before foam appeared on the Green. "The Chernobyl Ant was one of the first foam patterns we created," Emmett recalled. "We were trying to imitate the Mormon cricket, another terrestrial that occurs on the Green. The night we first tied it, Allan Woolley was over at my house, and we were having a few drinks and fooling around at the vise. Allan actually tied the fly; I named it. Later, I came up with the Mutant Ninja Cicada. Several guides on the river have contributed a lot in terms of cicada patterns, including Mark Forslund, Denny Breer, Carl 'Boomer' Stout, and Larry Tullis."

"You can't depend 100 percent on the cicadas coming off," Scott added. "Some years, the Green will only see a handful of bugs, and the trout aren't really on them. If we get a late spring snowstorm, it will kill the bugs off. Over the fifteen years that I've been on the river, we've gotten them ten years. Some years, though, we'll see two or three emergences. When this happens, we tend to see smaller bugs in mid to late May or early June, and the bigger bugs later in June. *Smaller,* of course, is relative. Mostly, we'll fish size 8s and 10s. There are times, however, when I'll go as large as a size 2. The timing of the emergence is dependent on the ground temperature; local wisdom is that they're likely to come out when the ground reaches 63 degrees. The cicadas also like it to be moist. The ideal scenario is a rainstorm followed by a warming trend. That's the trigger on the Green."

The Green River as we know it today was created with the completion of the Flaming Gorge Dam. The dam created a virtual spring creek, flowing crystal clear with temperatures in the mid-50s year-round. The fish that were planted here—browns initially, and later rainbows and cutthroats—have thrived to the point of ridiculousness on a rich diet of scuds. In some stretches, fish populations approach 14,000 fish per mile! Fishing is concentrated in the 30 miles of the Green below the dam, which is divided into three sections. Section A is enclosed by steep red rock canyons, and due to its large fish populations, sees the most angling pressure. Continuing downriver, the canyon opens up and fish numbers decrease. On the last 14 miles of the river, large browns—some over 20 inches—predominate, though fish populations decrease to about 1,000 fish per mile.

Scott described a typical day when the cicadas are on: "When the sun comes over the canyon walls and hits the trees, the bugs become active. The little ones make a clicking noise; the larger ones buzz. I like to see a little wind to blow the bugs onto the water. Cicadas are terrible fliers, and they land everywhere. Once the fish eat a few, they get pretty charged up."

Fishing the cicada emergence borders on fool fishing, and that's part of its joy. Working from a drift boat, one angler casts to the shore for the browns that tend to lie in wait in shallow pockets. Don't worry about being too close to shore; 18-inch fish will lie in spots where their backs are barely covered. The other angler can work the mid-river seam lines for often-explosive takes from the rainbows that prefer the faster water. At the peak of the cicada furor, anglers can realistically expect a fish to approach their fly on every second or third cast. It's not uncommon to set the hook on a rising fish and miss, only to have another fish take it after you've pulled the fly away. "There are times out there when you simply can't get down the river," Scott said. "The guy in the front of the boat will hook up, and I'll pull the boat over to help land the fish. By the time he has the fish in, the guy in back will have hooked up. Before we can get that fish in, the guy in front will have started casting and soon he'll have one on. I've had days with over eighty fish landed, and I wasn't even counting very hard."

A number of patterns have emerged to emulate the Green's cicadas. If the bugs are out in good numbers, they all will work. "I'm not sure that color matters too much," Scott added. "Many times when the bugs are on the surface, the fish are looking into the sun, and thus can't see the fly too clearly. The fly's silhouette is probably the most important thing, as the fish are looking for size and shape. There are some deer hair patterns that mimic cicada, but foam is much easier to work with and lasts multiple fish much better. A lot of guys will burn the foam around the edge to soften the curves. This makes the foam a little lighter, too, giving it a charcoal color. The bug's legs come out from the bottom of the body. I like to tie the legs that way, too."

Cicada COREY KRUITBOSCH

Green River Cicada BRIAN O'KEEFE

Green River Cicada

Hook: Gamakatsu S10, #8

Body: Black foam, cut about 2½ inches long and woman's finger wide

Thread: Black 6/0 or 3/0

Underwing: Purple UV Krystal Flash

Wing: White calf tail

Legs: Black/orange centipede rubber legs; 2 small diameter for inside legs, 2 medium diameter for outside legs

Superglue (to hold the fly on the hook and for extra weight)

Scott Barrus has been guiding since 1998 and now owns Spinner Fall Guide Service (www.spinnerfall .com). Scott was on the winning team of the Green River Single Fly Event in 2010 and 2011. When he's not at work, you might find him with his family, snowboarding, steelheading, traveling, or surfing.

Golden stones may play second fiddle to salmon flies on many systems, but they can frequently steal the show on Idaho's South Fork of the Snake. BRIAN O'KEEFE

Chapter 7

Golden Stoneflies

Elden Berrett

When it comes to stonefly hatches, golden stones seem forever destined to run second to their more flamboyant cousin, the salmon fly. Yet on many western rivers—even those where salmon flies come off in significant numbers—golden stones can steal the show.

"Golden stones are a phenomenal hatch on the South Fork and the Henry's Fork of the Snake," Elden Berrett began. "On the South Fork, it generally begins a week before the salmon fly hatch, somewhere around June 20 or 25. The golden stone emergence will make its way from the confluence of the Henry's and South Fork up 60 miles to the Palisades Dam; it won't get there until the middle of July. Many people don't even fish it; they think the goldens are early salmon flies. The goldens will back off a bit as you come into the heart of the salmon fly emergence, but a week or so after that hatch has petered out, the goldens will be back. If you fish along the grassy banks, you'll see millions of them; that's where they hide and mate. Some years you'll see them straight through November. Even in the midst of the salmon fly hatch, I'll use a golden stone pattern. The fish haven't seen a thousand of them, and something a little different can stand out."

Though their emergences sometimes occur simultaneously and the bugs are occasionally confused by anglers, salmon flies and golden stoneflies actually belong to different families. Goldens (*Hesperoperla pacifica* and *Calineuria californica*) reside in the Perlidae family, while salmon flies (*Pteronarcys californica*) are part of the Pteronarcyidae family. (*H. pacifica* is most prevalent in the Rockies.) Golden stones are on the whole slightly smaller than salmon flies, and a bit more muted in appearance. Before foam appeared on the market, most golden stone patterns were elk and deer hair spin-offs of the Stimulator or Sofa Pillow. With the advent of foam, Berrett's Golden Stone became a mainstay on the forks of the Snake—and many other rivers that boast golden stones.

Elden shared the tale of the pattern's evolution: "When tan foam came out, everyone was tying the foam straight to the hook. Eventually I saw one fly that had a little thread on the hook, and I wondered what it would be like if I tied a heavier body. I did one with yellow dubbing, and it worked surprisingly well. I kept working at it, though. I got my hands on some ginger-colored dubbing—the same color as the golden stone's body—and tied some bugs up. The next day, it was almost criminal. It didn't matter what water you fished, or how you fished it.

The trout were on it. I was getting forty to fifty fish a day; the best the other boats could manage was eight or ten fish. This went on for ten days. Finally, the guy I was guiding for—Lon Woodard, who owned Drifters—came to me and asked, 'What flies are you using?' 'Why?' I asked. 'I hear you've been catching fish in the frog water,' he replied. I gave him two of the flies. Lon came to me the next day. 'I will buy every fly you can tie,' he said. I told him I couldn't do it for cheap and he replied 'I don't care.' I was selling 'em for $2.50 apiece, this at a time when flies were going for $1 or $1.25. All the other guides were meeting me at the shop early in the morning and buying out my stock."

"One of the mistakes people make when they're tying golden stones is they make the profile too large," Elden added. "I think they're confusing the golden's behavior with how salmon flies act. Salmon flies flutter when they hit the water, struggling to get off the water. That creates a big profile. Goldens bounce on the water two or three times, and then they're either airborne or they don't make it up. If they don't make it up, they lay their wings down flat and try to swim to shore. There's no big fluttering wings; instead, you need a long, skinny profile."

If the profile of the golden stone can make or break you on the water, exactly how it's fished may matter less. "From the guide's standpoint, the golden stone hatch can be a gift," Elden continued, "as it can reward novice anglers who aren't accustomed to mending. When an angler doesn't mend properly, a belly develops, and the fly drags across the surface. This imitates that swimming motion and on some types of water, will entice strikes. Once the fish have had some pressure, though, dead drifting your fly along a bank with heavy water will be productive, as the fish stack in there waiting for naturals to come off the bushes. The monster fish will sometimes sit in back eddies eating naturals. A cast in there with a very slow strip and maybe a little rod tip movement—so the bug looks like it's swimming—will take some big fish."

Like any big bug on the surface, the golden stone arouses a fair amount of excitement for anglers—and, on occasion, some unwanted drama for guides. "A few years ago, I had a guy in the boat who owns a major pizza chain," Elden recalled. "He's a real fly-fishing addict, and on this occasion, he had a friend from the Midwest along, an avid bass fisherman. This fellow had never fished from a drift boat before, but he picked up the casting pretty quick. We were coming down the South Fork to a spot where a little creek dumps in. There's a seam there and a pocket of soft water where the fish always

Golden Stone BRIAN O'KEEFE

stack up. Right below this spot, the water picks up speed. There are some big rolling waves—nothing too technical, but you have to keep the boat straight and be on your toes. You get one shot in there, but you almost always get a take. I told the bass fisherman that there was a good spot coming, he should

Berrett's Golden Stone KEITH CARLSON

get ready to make a cast. He made a perfect presentation, and I told him to hold in, he was going to get a fish. Sure enough, a 16-inch cutthroat came up and ate the fly. When he set the hook, he stepped back with his left foot and then with his right. He went right over the side!

"A million things went through my brain, but I had the presence of mind to reach out and grab him and the rod. I tossed the rod into the boat and told him to hang on until we got through the fast water. As I got back on the oars, he worked his way to the back of the boat and crawled in.

"By the time I pulled over to the bank, he'd gotten to the front of the boat He picked up the rod and started reeling in the line, and the fish was still there. Soon he landed it. I removed the golden stone from the fish and we all sat silent for about a minute. Then we all busted out laughing."

Berrett's Golden Stone

Hook: Standard dry fly, #4–10

Tail: Brown goose biots

Body: Underwrap of floating poly yarn (for bulk and buoyancy) and ginger-colored dubbing

Hackle: Brown, palmered over the body

Wing: Tan foam, ⅛-inch-wide strip, cut to the length of the fly, tied at both ends over the body like that of a Chernobyl Ant. I use 2 pieces, one tied over top of the other, to give the fly more buoyancy.

Legs: Barred cream-colored medium round rubber legs

Indicator: Orange on one end and yellow on the other

When **Elden Berrett** isn't working at his full-time job, he's guiding the South Fork and Henry's Fork of the Snake. When he isn't guiding, he's fishing, which makes him a practical fly tier with patterns that have evolved from need. Elden is a signature tier for Idylwilde Flies; two of his flies have won the famous Jackson Hole One Fly and have caught the biggest fish in the event several times.

Though technically a tailwater fishery, the Missouri has been called the world's largest spring creek. GREG THOMAS

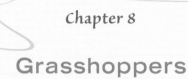

Chapter 8

Grasshoppers

Dave Bloom

Crisp morning temperatures give way to early afternoon warmth as the sun rises above the surrounding hills. A light breeze comes up, and the telltale sounds of stridulating orthoptera can be heard above the rustling of riverside grasses.

For anglers on Montana's Missouri River—and countless other rivers—this sound means one thing: It's hopper time!

Though he grew up in Great Falls and considers the Missouri his home river, one of Dave Bloom's earliest hopper memories came while fishing on the nearby Smith. "Growing up, I had a fishing mentor named Dick. Dick was a renaissance man; he was a successful surgeon, and when he decided he wanted to paint, he became a great painter. One time Dick took me over to the Smith. He fished a cane rod, and I remember him dropping Gartside Hoppers against the far bank and slaying brown trout. After seeing that, I pestered him to teach me how to tie those patterns."

Whether you're on a western tailwater, a Michigan freestone, or a Pennsylvania spring creek, grasshoppers fill an important slot. The large (and thankfully clumsy) insects seem to kick into their most active phase just as the summer's major hatches subside, and they're blown (or fall) into the water until the first hard freezes of the fall. Most active during the warmest times of the day, hoppers pose legitimate prey for hungry trout, providing afternoon surface fun when few mayflies are to be seen. This is certainly true on the river that's been dubbed "the Mighty Mo." Sometimes called the West's biggest spring creek, thanks to its clear water and myriad currents, the section of the Missouri (America's longest river) of greatest interest to fly fishers spills forth from Holter Dam and flows 42 miles to the town of Cascade. This stretch averages 4,000 browns and rainbows per mile, with fish averaging close to 17 inches. (Dave's biggest fish to date is a brown that went 32 inches and 15 pounds.) "It's truly one of the most amazing dry-fly fisheries I've ever experienced," Dave added. "The fish are big and they're also smart. It's a good test for any angler's skills."

For anyone fishing the Missouri these days, it may be hard to believe that at one time, floating the blue ribbon stretch could be a lonesome experience. "When I was growing up, there was no one else fishing on the Missouri. Your boat might be the only one out there. The fish, of course, are still here, and they're still interested in taking hoppers. At the very beginning of the grasshopper season, you can do well banging patterns against the banks. But as time goes on, the trick is finding spots where the fish haven't seen a hopper pattern—or fifty, or a thousand of them! For example, faster, skinny water." Some anglers are tempted to gurgle or skitter the fly across the water. For Dave, less movement is more. "I think a lot of people can overdo the action," he continued, "especially on a calm day on a calm stretch of river. In these circumstances, a little shiver of the fly as it dead drifts can close the deal, but I don't like more movement than that. On a windy day with choppy water, extra action doesn't hurt you."

Dave shared a few more bits of advice while hopper fishing on the Missouri: "First, pay attention to your approach. You only have one good shot, especially with bigger fish. Try to make sure the drift is right, and that you get the fly to the fish before it sees you. Second, if a fish rises to your hopper, let it turn; otherwise, you'll be singing the blues, with your hopper flying back over your shoulder as you set the hook on air." On some systems the hopper/dropper combination is a tried-and-true, best-of-both-worlds approach—the fish comes up to look at the hopper and takes the beadhead nymph on the way down. On the Missouri, Dave likes to keep it simple. "I hardly ever use a dropper out there. I think it screws the drift up, especially with the complex currents we have. If you want to catch a fish on a hopper, stick with the hopper and have your leader tuned up to a sturdier test so you can get a good drift. The fish will come."

Over the years, grasshoppers have inspired many memorable patterns—even before the advent of foam! Bigger Muddler Minnows were deployed as an early hopper imitation. Ed Shenk took the Muddler and adopted it to create the Letort Hopper. (Around the same time, in the late '50s, Ernie Schwiebert also developed a hopper of the same name; Ed's pattern became known as Shenk's Letort Hopper.) There's the Joe's Hopper (also known as the Michigan Hopper, and not created by Joe Brooks, though he fished it) and, of course, Dave's Hopper (the Dave being Dave Whitlock).

Though foam patterns have proliferated, Dave Bloom doesn't fish foam throughout the hopper season. "I find that foam patterns like Morrish's Hopper work really well some days," he said. "They look great and lots of people use them. However, once the fish take the cure, they get suspicious of them. When I guided on the South Fork of the Snake, Schroeder's Parachute Hopper was my go-to hopper pattern. The only problem was that the fly would sink in faster water. With Bloom's Parahopper, I took the Schroeder pattern as a jumping-off point and tied in a little life preserver of foam on the back. It keeps the fly up, but the body is still in the water like a natural. I like rubber legs for movement, so I added some in. I also glued in some pheasant wings, much the way that Dick showed me so many years ago. I'm not sure the fish see it, but it looks cool. Lastly, I like to tie in some CDC for the underwing. This traps some air bubbles and gives it a little more floatation. I think it helps to make the sale."

At the end of the day, fishing hoppers is about the big takes the big terrestrials inspire, and the potential for luring big fish to the top. "The takes are so varied," Dave enthused, "anywhere from slow and casual to ultraviolent. One that sticks in my mind would fall in the latter category. I have a client that I fish with each year who comes over from southern California. A few years ago, we were above a big tailout where I've hooked a lot of big

Grasshopper BRIAN O'KEEFE

Bloom's Parahopper KEITH CARLSON

browns. There's a spot where the water funnels into a V. The angler dropped his hopper into the apex of the V, and a brown took the fly so hard, it sprained his wrist. He was fishing the fly on 1X tippet. This is not a floppy-wristed guy; he's a cement worker. He didn't land that fish, but we came up with a name for it: the Wrist-Wrecker."

Bloom's Parahopper

Hook: Streamer (Dai Riki #710), #3X
Thread: Tan 6/0 UNI-Thread
Rib: Tan mono cord
Overbody: Tan foam
Body: Tan hare's ear dubbing
Underwing: Gray CDC
Wing: India hen back with head cement
Post: Pink poly yarn
Hackle: 1 brown and 1 grizzly hackle
Legs: Brown rubberlegs
Head: Hare's ear dubbing

Dave Bloom began his guiding career (www.bloom
outfitting.com) in 1985 on the Henry's Fork, the
South Fork of the Snake, and the Teton in Idaho. In
1995 he moved to the famous Montana rivers around
Yellowstone Park. Since 2001 Dave has focused his
guiding efforts on the Missouri in the summer and in
southern Chile and Argentina in the winter.

The Dave behind Dave's Hopper

Before there were Chubby Chernobyls—heck, before foam had become a regular ingredient in the fly tier's arsenal—there was the Dave's Hopper. It wasn't the first grasshopper fly ever tied, but it's certainly one of the most enduring patterns, and has served as a template for many creations that have come along since it first appeared in the early 1960s.

Its creator, of course, is Dave Whitlock, one of fly fishing's most recognized and accomplished personalities. Dave is not only a gifted fly tier, but a fine writer (his books include *Trout and Their Food, Dave Whitlock's Guide to Aquatic Trout Foods,* the *L.L. Bean Fly Fishing Handbook,* the *L.L. Bean Bass Fly Fishing Handbook,* and *Imitating and Fishing Natural Fish Foods* for Lefty's Little Library), illustrator, and educator. An Oklahoma native, Dave spent many years in Arkansas's Ozark Mountains on the White River with his wife, Emily, who hails from the Razorback State. The Whitlocks recently returned to Oklahoma, where there's a lot more good fly fishing than most would think.

In recent years, Dave has become involved in trout stream design and restoration. One Whitlock innovation you may not know about is the Whitlock-Vibert Box System, an in-stream salmonid egg incubator and nursery device. The Whitlock-Vibert Box system is in use throughout the world, in conjunction with the Federation of Fly Fishers.

Kamchatka's Sedanka River fosters rich insect life, including green drakes. RYAN PETERSON

Chapter 9

Green Drakes (Western)

Mike Mercer

"I still vividly remember my first green drake hatch," Mike Mercer began. "I was 12 or 13 and had gone on a fishing outing up to Hat Creek with the Chico [California] Fly Fisher's club. It was the early '70s, and Hat Creek had just recently been made fly fishing only. Some of the guys on the trip said, 'You might see some salmon flies.' I had started fly fishing a year earlier; I didn't know a ton, but knew a little. I went off on my own and fished nymphs and found some beautiful rainbows, 13 to 16 inches—much bigger than anything I'd ever caught in Chico. I was thinking it was the most amazing river I'd ever seen.

"The group gathered just at dark to get ready to go home, but then we noticed loud noises coming from out on the river, like someone was chucking bowling balls. 'Maybe it's a green drake hatch!' one of the guys said. Sure enough, it was. I could just make out these goliath bugs riding the current next to shore. They were size 10 or 12—huge to me, as I was used to little mayflies. Someone handed me a fly and said, 'Tie this on!' I did and cast it out there into the darkness. Soon I had a fish."

With the exception of a few systems that harbor *Hexagenia limbata,* green drakes are the largest mayflies anglers can expect to encounter on western waters. On some rivers, emergences (and especially spinner falls) occur close to dusk; on others, it's the early afternoon. One thing that's fairly certain about green drakes is their uncertainty. "It's a pretty mercurial hatch," Mike continued. "Some evenings, there will be a light occurrence and just a few fish up on top. Other nights, the fish will be just stupid for fifteen or twenty minutes; you can catch fish as quickly as you can cast. You might go ten days between big hatches, but the guesswork and anticipation is part of the draw. On Hat Creek in June, when you hear that first cannonball 'whoosh' and look up against the sky and don't see salmon flies, you know it's green drakes."

Mike recalled the time he made the transition from wide-eyed adolescent angler to young guide trying to make his mark . . . and the role that green drakes played. "When you're a guide, you have to produce," Mike said. "You're always thinking, 'How can I buy my client a few more fish?' If you're not constantly learning when you're guiding, things aren't going to end well. At that time, it was legal to seine the rivers, and I started doing so. In the riffles on Hat Creek, I found lots of salmon fly nymphs, some golden stone nymphs, and lots of green drake nymphs. The latter were big bulky bugs, size 10 or 12, and very robust, with big thoraxes and heavy legs. The gills were very pronounced, and they fluttered in the water. As a tier,

I wanted to capture that heavy thorax and distinct gill structure. This led to the birth of the Poxyback Green Drake Nymph. I dabbed a little epoxy on the back to capture the shininess of the natural's wing case as it fills with gasses to blow the wings out. I felt like getting the gills right was the other main challenge. I hit on using filoplumes. I tied them on top of the dubbed abdomen, and the undulating feathers really captured the movement of the gills. I've gotten great feedback on that fly over the years. I've heard of people fishing it all over the country, even in places where there aren't any green drakes. Something about it evokes life, and some anglers even use it as a searching pattern."

Fast-forward another ten years; Mike was then building a reputation as a fly designer and helping to hold down the fort at the Fly Shop in Redding. "I wasn't guiding anymore, but I still got out to fish a lot," Mike continued, "and I wasn't very pleased with the green drake adults we had in the shop. They seemed rudimentary, and didn't float very well. When you have a hatch that you can only fish for twenty minutes, you want a fly that won't sink and will stand up to a few maulings. I'd seen these extended-body foam-body flies that a guy up in Oregon was tying out of packing crate foam. They were called Bunse Duns, and I admired the flies. I wanted to try to come up with something like that. I started using Larva Lace dry-fly foam and extended the foam abdomen off of a short-shank hook. I tied them with a bright-colored upwing, which is really easy to see in any kind of light. The parachute hackle pushes the downwings into the water film, which is important. I can use the fly—the Foam Profile Dun—for both spinners and duns. It's durable and floats high; as long as your tippet holds, it's good for an evening's fishing. It's been my best pattern for green drakes on Hat Creek, and that's saying something, as Hat Creek can be hard."

Green drakes are not limited to North America in their distribution. Mike has seen them as far afield as the Kamchatka Peninsula in eastern Russia, where they once rescued a morning's fishing. "Of all the places we send people in Kamchatka, the Sedanka is the only true spring creek," Mike explained. "It has daily hatches; sometimes small bugs, sometimes larger. It's really a gorgeous stream. It pops out of a big lava flow, creates a spring pool or two, and then flows for miles down a gentle gradient channel. You can hop across it at the headwaters, and then it becomes much larger. At the uppermost camp, I had some wonderful mouse fishing. Then, further downriver, we came to a huge salmon spawning redd. It was probably 20 yards wide and 50 yards long, just crammed with sockeye and cherry salmon. Sometimes I could make out the rainbows amongst the salmon. 'I'm gonna skate a mouse through there and pick 'em off like ripe fruit!' I thought. But I hadn't counted on the salmon being so aggressive; I couldn't get a mouse halfway through a swing without a sockeye or cherry eating it. I thought about streamers, but knew I'd only snag salmon. The water was loaded with Dolly Varden, too, so an egg pattern was out—the Dollies will beat a rainbow to an egg pattern, every time.

"As I stood there puzzling, I recalled that there had been a green drake hatch the day before. There were no duns on the surface at this moment, but I thought perhaps a drake nymph might work. I knew the salmon and the Dollies wouldn't be as likely to eat

Green Drake BRIAN O'KEEFE

Mike Mercer's Foam Profile Dun BRIAN O'KEEFE

it. So I put on my Poxyback and a little yarn indicator and cleaned up. Every three or four casts, I'd hook a big rainbow—18 inches, 21 inches, 23 inches. I went through and caught thirty big fish in two hours; it was like picking up Easter eggs on the lawn, almost too easy. As it turned out, this was the only occasion on that trip that I had to nymph. Yet it was one of the more memorable fishing experiences of my life. I had been able to read the situation and come up with a solution. It was as satisfying as fishing a dry. Every time I drifted the nymph and indicator through a new little streambed depression, I was thinking, 'It's gonna get eaten, it's gonna get eaten!'"

Mike Mercer's Foam Profile Dun

Hook: TFS 2500, #10 or #12

Thread: Olive 8/0 UNI-Thread

Tail: Neck hackle, grizzly dyed olive-feather stems, stripped

Extended abdomen: Midge tubing, yellow fly foam, folded over a needle; colored with a Prismatic permanent marker, olive on top, light tan on bottom

Rib: Olive 8/0 UNI-Thread

Wing post: Orange and yellow McFlylon Poly Yarn

Downwings: Dun Z-Yarn

Thorax: Mercer's Buggy Nymph Dubbing, micro mayfly

Parachute hackle: Saddle hackle, grizzly dyed olive

Mike Mercer manages the Fly Shop's Alaska Travel operations. He has fished many of the world's

fly-fishing dream destinations and is recognized as one of the West's leading fly designers. Mike is an Umpqua Feather Merchants contract tier, and many of his fly designs are being used successfully around the world.

In a state not known for its trout streams, Connecticut's Farmington River has emerged as one of New England's premier dry-fly waters. THOMAS BARANOWSKY

Chris Conaty

"I grew up in Southington, Connecticut, in the middle of the state," Chris Conaty began. "We had a crab apple tree in our backyard. When that tree was in flower, you could be certain that the Hendrickson hatch was in progress on the West Branch of the Farmington River, about 30 miles to the north. You could call up one of the tackle shops up there, and they'd confirm it. That crab apple tree was an incredibly accurate biologic indicator."

For anglers in the East and Midwest, the appearance of *Ephemerella subvaria* is a sign that spring—and more importantly, trout season—has arrived in earnest. This mayfly—in many venues, the first good hatch of the season—is encountered from southern Appalachia through Pennsylvania, west to Michigan and Wisconsin, and throughout New England as far north as Maine. In the southern part of its range, it might occur as soon as early April; farther north, it appears toward the middle or latter part of May. *Ephemerella subvaria* can be found in a variety of habitats, from low-gradient, gravel-bottomed streams to the pocket water of fast-flowing mountain freestones. While fish may show interest in the spinner fall, the most intense action generally focuses around the duns, which float like armadas of grayish-sailed sloops through troubled waters punctuated by dimpled snouts. (Males frequently have a more reddish tint and are sometimes emulated with another old standby, the Red Quill.)

The Hendrickson pattern has a provenance tied closely to the history of American fly fishing. The story goes that one Albert Everett Hendrickson of Scarsdale, New York, wrote a letter to Roy Steenrod, a friend and acolyte of the legendary Theodore Gordon, about creating a pattern to match the dark up-winged dun of the *Ephemerella subvaria*. Steenrod, who frequented the Beaverkill, Esopus, Nerversink, and Willowemoc Creeks in the Catskills, invited Hendrickson to come fishing. One day in 1916, while fishing the Beaverkill below the junction pool at the town of Roscoe, a prodigious hatch of the mayflies came off. Steenrod plucked one of the insects off the water, and over lunch, tied a fly to match. It proved to be a winner, and the pair caught fish after fish that day and the next. On the third day, the matter of naming the fly arose, and Steenrod graciously named it after his new fishing friend. In a twist that speaks to the ubiquity of the pattern and its incredible productivity, we now refer to the insect—not just the fly—as the Hendrickson!

"My dad taught school in Southington," Chris continued. "Each spring—always in early May, and frequently on May 3, just after the crab apple had blossomed—there was a day during the week when he'd tell me, 'We're going to miss school tomorrow and go to the river.' He'd call in sick, he'd call me in sick, and we'd drive up early enough to scout the river and stake out a spot well before the hatch began, which was usually around 1 p.m. I was 8 or 9 the first time I went along. We'd sit on the water, waiting for the first bugs to emerge and the first fish to rise. People had good etiquette and would ask if you were resting the water. If you said you were, they would move on. I'd stand next to my dad; he'd cast, get a rise, set the hook, and hand me the rod. They were all hatchery fish in those days on the Farmington, probably planted a few days before. I remember that sometimes for the first few days the Hendricksons came off, the fish wouldn't rise in good numbers. I believe it was because the flies—or, for that matter, the naturals—didn't look like the pellets they were used to. After a while, they caught on. Some days, it was truly magnificent, with fish rising every 3 or 4 feet across the pool. The hatch might go on an hour or two. We'd fish swimming nymphs and Comparaduns, which had just come out around this time. They were revolutionary, and became the go-to pattern. As I grew into a teenager, I tried tying Comparaduns. I found it difficult to fan out the hair so the fly would land on the water upright. My flies would often fall on their side. To my surprise, the fish would eat these even better. This was a lesson.

"My dad had a group of fishing buddies, four other teachers from town, and on those midweek trips, we'd often meet one or two of those guys up there. They'd have skipped school as well. I have a picture of my dad with one of his buddies from that time on the banks of the Farmington. They both have long '70s sideburns, longer hair, checked flannel shirts, and canvas chest waders. They each have a can of Schlitz beer in their hand, and there's a Styrofoam cooler in the background. They're waiting for the hatch to come off. Every time I look at that photo, I can smell the crab apple tree in my old backyard and can see the heads poking up in that pool."

Hendrickson DAVID SKOK

Hendrickson BRIAN O'KEEFE

In a state not exactly celebrated for its trout fisheries, the West Branch of the Farmington is a bright spot. Year-round cold water from the Colebrook and West Branch reservoirs and aggressive management practices have made the Farmington southern New England's premier trout fishery. The Farmington is still supplemented with hatchery fish, though a smattering of browns and rainbows—both wild fish and holdovers—can push the 20-inch mark.

"I've been living out in the western United States for almost thirty years," Chris continued, "but I try to get back to Connecticut each year to fish with my dad. I can only get away in August, so I miss the Hendrickson hatch. But we fish the same water, and I'm brought back to the crab apple tree and my dad and his fishing buddies in their checked shirts. It's not the kind of fishing I'd spend a minute on out west, but the memories run deep."

Chris Conaty is director of product development for Portland, Oregon–based Idylwilde Flies. Chris has taught climbing and wilderness travel, and he has designed rope courses for various organizations around the world. He prefers fishing for summer steelhead with a dry line and beautiful hair-wing fly to all other fishy pursuits these days.

Your reward for wading through mosquito-infested swamps and casting pinkie-size flies blindly into the night. BRIAN O'KEEFE

Kelly Galloup

There are "blanket" hatches. And then there are hatches that more resemble the kind of foot-deep insulation that one might find lining a research facility at McMurdo Station.

Welcome to the Hex hatch!

"Nothing in fly fishing even comes close to the *Hexagenia limbata* spinner fall," Kelly Galloup declared. "I don't care how big a salmon fly or Mother's Day caddis or Hendrickson hatch you've seen, it doesn't compare to the Hexagenia in pure biomass. In a big Hex emergence, the bugs can literally cover the river, like a whiteout. You can't cast through it; you can't see rise forms. It's game over. I have a photo somewhere of a time that I set my rod and reel on a platform of bugs on the water, and the bugs were so thick that they suspended the outfit over the water. Large Hex hatches in the Great Lakes region are so vast, they show up on Doppler radar. The clouds of bugs are miles long. There are times when the streets need to be plowed to prevent piles of mayflies from becoming a road hazard. In the early summer, the freighters and racing sloops out in Lake Michigan run without lights. If you run a light, the boat will end up covered with 2 inches of Hexes.

"I have a friend who built a cabin on Skeegamog Lake. The house went up in the winter and spring. He put in a mercury light in his yard, though he noticed that none of his neighbors seemed to have lights. He found out why. The first summer, he woke up to find his wife's white Cadillac covered with 6 inches of bugs, and a 3-inch layer of bugs on the house. He had to have the exterior of the house spray washed and then vacuumed, and some of it repainted. It cost $8,000. He took the light down."

The biblical proportions of *Hexagenia limbata* occurrences would be little more than a curious phenomenon of the insect kingdom if it weren't for the fact that these big bugs—North America's second largest mayfly—bring the truly big fish up to the surface. "I don't know if there's a better opportunity available to catch big fish on dry flies," Kelly continued. "One season I had nine fish over 28 inches on dries, including one night when I landed three such fish—a brown, an Atlantic salmon, and a steelhead—without moving a step. The anticipation the hatch fosters is tremendous. It's like stepping onto a productive tarpon flat. It doesn't always happen, but if it does, it's special. To hear an 8- or 10-pound fish eat a dry fly a rod length or two away in the dark—that's something unique in this business."

Hexagenia limbata is one of North America's most widely dispersed mayflies, found in lakes and slow-moving streams from Florida to Washington. If a system has a soft, silty substrate rich in calcium carbonate, chances are good that it will support Hexagenia—sometimes in densities approaching 500 nymphs per square foot. After a year or two in a burrow, they build in the silt, the nymphs emerge, reach the surface, and molt. The mayflies mate in the sky, and the females float down to the water, release their eggs, and expire. "In Michigan, it seems like all the drainages have a Hex hatch, from famed rivers like the Manistee, Pere Marquette, and Au Sable down to little creeks that empty into Lake Michigan or Lake Huron," Kelly continued. "In my later years in Michigan, I'd spend more and more time in estuaries where creeks dumped into Lake Michigan, as the big lake fish key in on the emergence."

Like the mania that grips steelhead devotees in the Pacific Northwest, the Hex hatch engenders a certain twisted passion among Great Lakes anglers. "My record in fishing is sixty-one straight nights," Kelly confessed. "Most of the hatches last ten to fifteen days, and begin in the southern part of Michigan where the water's a little warmer. You can chase the hatches from the bottom of the state to the top. I can't tell you how many guys I've met who have dogs named Hex." That being said, the Hex's appeal is a nuanced one. For starters, the action unfolds in the dead of night.

"To many anglers, having lights out is either frightening, boring, or both," Kelly continued. "So many dry-fly guys want to see the rise, the take. That's cool, but it's not Hex fishing. You can fish a Hex pattern all day and you'll do nothing. The spinner falls and the big fish are out at night. You're often in a cedar swamp, which makes it even darker. I don't use a light, as you'd be covered in bugs in an instant. We'd wear hoodies with the hood drawn down tight to keep the bugs at bay. On big nights, you have to keep your face down so you don't eat them.

"The night begins by posting up on a hole. You're going to fish to a specific spot—ideally a funneling area where the bugs drift close by—and stand there until the night is over. Then you wait. The nights when a hatch occurs follow a pattern: At 10:30, the spinners begin to hit the water; by 10:45, every fish in the pool—mostly the smaller guys—are feeding; by 11, you'll hear the occasional big splash. That means the big fish have arrived. At that point, the smaller ones usually leave. On a good night, you can hear the Hexes before you can see them. It's like a small wind blowing over your head. When they hit the water, you're fishing by sound. You try to find the cadence the fish are feeding at and cast accordingly. Oftentimes, you're only casting two rod lengths' worth of line. The biggest fish I ever caught during the Hex hatch—a brown just shy of 3 feet—wasn't even a rod length of line away!"

Kelly developed the first commercially produced Hexagenia pattern—the Troutsman Hex—around 1980. "We had a dun at first with an upright wing," he explained, "and we also took the same pattern and spent-winged it, for the spinner. I like to palmer the body on my spinners. I think the palmering represents the vibration the naturals have on the water, the quivering, like little shock waves. The patterns were generally tied on size 4s." Considering that the trout are feeding in utter darkness, one must ask, how much does the pattern really matter? "I can say one thing unequivocally," Kelly added. "Using a brighter fly on a brighter night—say with yellow bucktail and white wings—and a darker fly on a darker night with purple wings improves your catch rates. Though that being said, I've seen brighter flies taken on dark nights and vice versa."

As alluded above, fishing the Hex spinner fall isn't for everyone. "I guided the Hex hatch on some of Michigan's fabled waters for twenty years," Kelly added. "That's where I learned to outfit my rods with reels that only went in one direction. In the heat of the moment, in the darkness, some clients would become disoriented and reel in the wrong direction. I had a few occasions when we were fishing from a boat, and anglers would get vertigo and fall out of the boat. On one occasion, I was guiding for a group of anglers at Ranch Rudolph, which hosts an annual fly-fishing school for Trout

Hexagenias BRIAN O'KEEFE

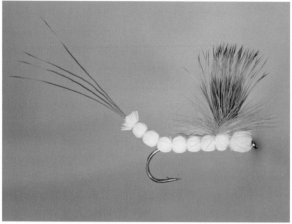

Extended Body Hexagenia Drake BRIAN O'KEEFE

Unlimited. I had one of the lead casting instructors as a client, and we postholed in a spot around 10:30. On schedule, every fish in the ditch began to feed, mostly little guys. Then we could hear a 10-pounder about 10 feet away from us. The fish actually splashed us when it ate. My client had an anxiety attack. I had to pull him off the river, and he later drove himself to the hospital. This is a guy that had fished all over the world.

"There have been a handful of occasions over the years when I've seen the Hexagenia spinner fall come much earlier in the evening, around 7 p.m. I'm talking about six times in thirty or forty years. If it happened more frequently, I'm not sure there would be any trout junkies left fishing in the Rockies."

Troutsman Hex

Hook: TMC 5212, #6 or #8

Thread: Yellow 6/0 UNI-Thread

Tail: Moose mane fibers

Body: 1 brown, 1 yellow clump of deer hair

Rib: Tying thread

Wing: Calf tail

Hackle: 1 grizzly, 1 brown dry-fly hackle

Kelly Galloup opened his fly shop, the Troutsman, in Traverse City, Michigan, in 1981. In 2002 he purchased the Slide Inn on the Madison River in Montana with his wife, Penny. An author and lecturer, his books include *Modern Streamers for Trophy Trout* (co-authored with Bob Linsenman) and *Cripples and Spinners.* He is editor-at-large for *Fly Fisherman* magazine and one of the hosts of *Fly Fish TV* (www.flyfishtv.com) on the Versus network.

The appearance of March browns on western rivers (like the McKenzie in Oregon) are a harbinger of better weather to come.
BRIAN O'KEEFE

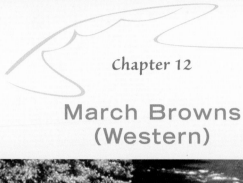

Chris Daughters

Nothing (short of a free trip to Andros Island in the Bahamas) warms a winter steelheader's heart like the promise of a floating line, a 4-weight rod, and a fly that won't knock you unconscious should it veer off target on your forward cast. "If I'm out on the coast steelheading in February and see a March brown on the river," Chris Daughters offered, "I can be pretty sure that we're just two to four weeks away from seeing their emergence on the McKenzie."

Like the Skwala stonefly in western Montana, the western March brown (*Rhithrogena morrisoni*) is the harbinger of warmer spring days and steady dry-fly action in the western United States. (Its namesake in the Midwest and East is another important emergence, but a different mayfly, *Maccaffertium vicarium*.) On some western systems, March browns are the first decent-size mayfly of the year; on others, they are preceded by the aforementioned Skwala. "It's hard to get a consensus on when the hatch will come off on the McKenzie," Chris continued. "The hopeful are beginning to look for bugs in the middle of February. During warmer, low-water winters, it could happen that early. Early March to mid-April is a pretty safe bet. When it does happen, the fish are interested, especially the rainbows, which have recently finished spawning.

While there are Baetis and some winter stones around, it's my experience that the bigger rainbows won't move to them. For the March browns, however, the fish will set up in the shallower riffles and the runs below them where the nymphs emerge."

The McKenzie is one of Oregon's best-known rivers, revered for its beauty, accessibility, and rich angling history. With its headwaters at Clear Lake in the Cascade Range, the McKenzie flows some 90 miles west through the heavily forested Cascade foothills and pastoral McKenzie Valley until it merges with the Willamette just north of the city of Eugene. The upper stretches offer abundant whitewater challenges and excellent angling for native redband rainbow trout. This is tricky fishing, requiring fast, accurate casts to small pockets of calm water behind rocks, but the rewards for the skilled can be great. Below Leaburg Dam the river flattens out, opening into a more leisurely series of pools and riffles. Fish here are a mix of cutthroats and rainbows.

If you didn't know the McKenzie for its trout, you might know it for the drift boats it inspired. The dories New Englanders brought west were poorly designed for use on Oregon's shallow whitewater rivers. Soon the boats began evolving: Deep displacement hulls were replaced with wide, flat bottoms, and the size of the rocker (the upward curve of boat bottom toward bow and stern) was increased. These lighter boats soon displaced the dories on rivers like the McKenzie, where fishermen—and more importantly, fishing guides—recognized the value of their greater maneuverability. In the 1930s a woodworker named Torkel Gudmund "Tom" Kaarhus began building boats in Eugene; he soon designed and crafted the first square-ended McKenzie drift boat. One of Kaarhus's associates, a fishing guide named Woodie Hindman, added his own touches to the design, which became widely available by the 1940s. Through the combined efforts of Kaarhus and Hindman, the drift boat that's used by thousands of guides and recreational anglers was born.

The March brown hatch is a most gentlemanly occurrence, beginning late morning and ending mid-afternoon. "If you work in Eugene, it's very conceivable to build your lunch hour around the emergence," Chris continued. "You can bail out of work around 11:30 and sneak back in by 2, and you'll have caught the best of it. In the early part of our guiding season, we tailor trips around the hatch, putting in at 10 and taking out by 4. We'll usually start out nymphing with a two-fly rig, with a Hare's Ear or Mega Prince on the point. When we start to see a little activity on the softer edges, we'll swing soft hackles. Once there's consistent bug activity, I like to find a riffle and set up below it.

March browns are notorious clingers, and the nymphs live in the faster water. I came up with a Klinkhammer March Brown pattern that seems to work well when the fish are on top, though Comparaduns are effective, too. [Klinkhammer-style patterns, developed by Hans van Klinken, are tied so the thorax/abdomen of the fly rests below the surface film. Popular in Europe, they're seeing increasing use in North America.]When the March browns are coming off, you're not going to find them throughout the river. If you're in the wrong type of water, you're out of it. I like to drift when I'm fishing this hatch, as you have a better opportunity of finding a spot where it's happening. Even then, there's the chance that in the fifteen minutes it might take to get to the next spot, the emergence could wane. Sometime it's just better to sit on a spot that's been productive in the past and wait it out."

Chris recalled a drizzly March day on the McKenzie in 2010 when the planets were properly aligned. "I decided to do an exploratory trip with a couple of guys from the shop to see if the bugs were out, and they were. In fact, there were fish swirling right at the boat ramp. They were taking emergers right below the surface, and soft hackles did the trick. We were fishing two-fly rigs, and between us we had a number of double hookups—the fish were that grabby. Pretty soon, big, dark brown sailboats were coming down the river, close to size 10. Fish begin popping all around. We anchored on a flat that fishes consistently well. There are lots of cutthroats in this section, and many were rising. Something much larger blew up just off the right edge of the boat. I didn't have quite the right angle to make a good presentation from the front of the boat, so I handed the rod to Ethan Nickel, who was on the oars. I had a dry on top and a soft hackle tied off the back—fortunately it was on 8-pound test. Ethan made a little reach cast and the fish swirled on the soft hackle, took it just like a dry. The other fellow in the boat, Matt Stansberry, described the fish as 'a red flash that looked like a two-liter Coke bottle.' The fish ran all around the flat, and finally Ethan got it to

March Brown COLIN ARCHER

March Brown BRIAN O'KEEFE

hand after I'd beached the boat. It looked more like a steelhead in size.

"I didn't see a fish that good the rest of the season."

Klinkhammer-Style March Brown
Hook: TMC 2488, #12
Thread: Black 6/0 UNI-Thread
Tail: Lemon wood duck
Rib: Fine copper wire
Dubbing: Australian possum
Hackle: Brown
Post: Neon-colored Antron

Chris Daughters began fly fishing at age 10, taking classes at the Caddis Fly Shop (www.caddisflyshop .com) in Eugene, Oregon. He put himself through college working at the shop and guiding; in 1996 he purchased the shop. Chris still guides over a hundred days a year.

The Netherland's Greatest Fly Tier?

You may not associate the Netherlands with fly fishing, yet this watery nation is home to one of Europe's most influential fly tiers—Hans van Klinken. Hans has brought the world the Klinkhamer Special—a pattern that copies an emerging caddis, and which uses the fly's abdomen to poke through the surface. Since it was introduced in the early 1980s, the Klinkhamer concept has been adopted for imitations of many other insects.

Much of Han's early fishing transpired in Scandinavia. He caught his first Atlantic salmon in Norway at age thirteen using a spoon, though he soon transitioned to fly tackle. By age sixteen, he was traveling alone to Norway, Sweden, and Finland to fish; he took his first fly-caught salmon and sea trout on dries! He took up fly tying in 1976, and a few years later began improvising his own patterns—mostly dries, including early prototypes of what would become known as the Klinkhamer special.

Over the years, Hans has made fly fishing and tying a more and more essential part of his life. Since 1986, Hans has participated at many fly-tying shows and given lectures, classes, and workshops around the world. He's contributed to many books, including *The World's Best Trout Flies* by John Roberts and *The Complete Book of the Grayling* by Ron Broughton. Though he may always be best known for the fly that bears part of his name, other patterns—including the Caseless Caddis, the Reemerger, and the Once and Away—reside in thousands of fly boxes.

Anglers on New Mexico's San Juan River may not always see midges, but you can be sure the fish do. AARON OTTO

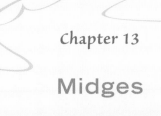

Rick Takahashi

Midges are the 97-pound weakling, the Rodney Dangerfield, of trout food. They're easily dismissed, under-respected, and frequently overlooked. Or perhaps one should say *underlooked,* as anglers of a certain age may find it challenging to locate midges on the surface, let alone attach an imitation to a gossamer-like filament of 7X tippet.

"I was on the San Juan River in New Mexico over twenty years ago," Rick Takahashi began. "At that time the river was quite healthy, and you could see the fish in the water. They were taking *something.* I tried all the patterns that usually worked in my native Colorado, and the fish ignored them. There were a few anglers downstream that were doing quite well. I finally asked them what they were using, and they showed me their midges. I used to think that midges were just a smaller version of an Adams or Black Gnat—though the smallest I'd consider then was size 18. These flies were *really* small—size 22 to 26. I decided I'd better learn more about midges."

Midges are members of the family of Chironomidae, two-winged insects that include blackflies and mosquitoes. They are present in significant numbers in still and running water systems across North America. Trout are generally most interested in midges in lower-gradient rivers and still waters (or slower pools/ sloughs in faster-flowing streams), where minimal effort is required to capture these little morsels. Rick has come to consider midges an important main course in the trout's diet. "One of the amazing constants about midges is that they are present nearly all year long in both still waters and rivers and streams," he continued. "As long as you have open waters to fish, you can almost be certain that midges will be active during the day. They continue to hatch in great numbers nearly every day of the year. Other food forms such as mayflies, caddis, and terrestrials in moving water and damsels, mayflies, and crayfish in lakes may deter fish from selectively feeding on midges. But be assured, even though other food forms are present, trout will ingest whatever food form is easily obtained and gives it the most protein value."

Where some hatches may be limited to a few weeks of the year (and even then, may require the angler to bide his or her time), midges are available almost all the time. "I'm always prepared to fish midges," Rick said, "though that doesn't mean I always will. I carry a wide variety of midge larva and pupa patterns with me in sizes 16 to 26 for rivers and sizes 10 to 18 for still waters. I do carry a small selection of midge adults and cluster-type patterns but find that the pupa stage of the midge life cycle is the most important developmental phase. In the spring when runoff is occurring, I'll fish lakes in northern Colorado with larger midge pupa patterns. As the rivers recede to more normal levels, I'll return to the moving water and fish a two-fly nymph rig with a midge pupa as one of the patterns. I like to tie the pupa 10 to 12 inches off the eye of the point fly, so it's in the fish's cone of vision. The pupa seems to be the most productive stage of the bug to fish. They're present in such abundance and are easy for the fish to capture, so they're worthwhile for the fish to pursue, even though they're small. Fish will stay in one feeding lane with their mouths open, waiting for the pupae to flow down. From midsummer to September, fish will focus on other insects, but as fall progresses, I'll again add midges to my strategies and fish them until ice and freezing cold weather no longer allow access to open water."

Some of Rick's most memorable midge fly-fishing experiences have occurred on the San Juan. "I had fished one of my favorite runs on the river located above the Kiddie Hole and the bottom part of the Upper Flats [both part of the catch-and-release section below the Navajo Dam]," Rick recalled. "I had often fished this section of the river with fairly good success, but I always encountered a pod of about fifteen to twenty rather large rainbows rising to midge adults back in some very slack water. I tried every pattern in my fly box, including some size 24 to 26 Parachute Adams. The fish would come up to the fly, inspect it, and slowly turn away; needless to say, I was very frustrated. One night I had a dream about a fly pattern that represented a midge adult. In my dream I designed a fly that had a wing of black Antron yarn, a tail of black Krystal Flash, a body of black porcupine guard quill, and a black hackle. It was such a powerful dream that it woke me up. I went to my vise, sat down, and tied a half dozen of these flies.

"I woke up early that morning, too excited about this pattern to sleep. I got dressed and headed down to the river, and was on the water by 7 a.m. The sun had been up for about an hour, and the river had a fairly dense low-lying fog as I waded into position. I approached the area where I had seen the fish the previous day. To my surprise, I could hear the gentle slurping noises of the fish feeding on the midge adults before I actually spotted the fish. I positioned myself above and slightly to the side of the fish, tied on the new pattern, crimped down the barb, and applied some floatant. I cast about 4 feet above the first rising fish I saw and adjusted the drift of the fly into its feeding lane. The sun had started to burn off the fog, and I then could clearly see about fifteen fish working the surface. As the fly drifted towards the first fish, I couldn't really make it out beyond the black Antron wing; I had tied the wing in a vertical position about two times the length of a regular wing. The trout rose confidently to the fly and took the new midge pattern on the first drift. I proceeded to catch all fifteen of the fish by 9 a.m. I thought that it doesn't get any better than this, and I thanked the Lord for giving me the vision of the pattern the night before!"

Rick related another tale from the San Juan that sheds light on his sharing nature: "I was on a slow braided stretch directly above the Kiddie Pool. I had found a stretch of water that had a deeper channel running through the middle of the run. I was hooking fish on a consistent basis when an elderly

Midge AARON OTTO

couple entered the upper section of this run. They saw me catching fish and decided to sit on a large rock and watch me fish. I greeted them and they said they hoped they weren't bothering me but wanted to watch me fish. I asked them if they wanted to get a closer look, and they said no, it was my run. They reported that they had fished the day before without much luck, and they hoped to get a hint or two from watching me. I told them to wade over to where I was and I'd show them the patterns, the setup, and how I fished the run. I gave them a dozen of the patterns I was fishing, showed them how to rig up, demonstrated how to control their drift, and told them to

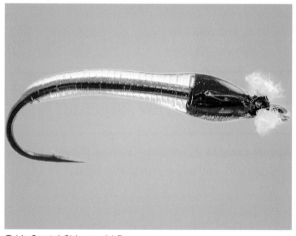

Tak's Crystal Chironomid Pupa COURTESY UMPQUA FEATHER MERCHANTS

start fishing. They asked, 'Don't you want to fish anymore?' I said, 'No, I've caught more than my share of fish.'

"I stepped back and watched them fish for a few minutes. I said my good-byes, and as I left the river, I could hear the woman say, "Oh, honey, I just caught a fish!' and they were both laughing and enjoying their success. I left the river with a big smile. Hearing them having fun and catching fish was way better than fishing."

Tak's Crystal Chironomid Pupa
Hook: TMC 2302 or 2312, or Daiichi 1260, #12–16
Thread: White 8/0 UNI-Thread
Body: .05mm clear Stretch Magic Lace*
Wing pads: Brown biot
Wing case: Pearl tinsel
Gils: Oral-B Ultra Dental Floss
Thorax: Tying thread colored with markers to match naturals
Colors: Light olive, tan, sand, light gray, cream

*Cover body with Loctite Instant Adhesive, UV Fly Finish, UV Knot Sense, or Hard As Nails.

Rick Takahashi recently published his first book, *Modern Midges: Tying and Fishing the World's Most Effective Pattern,* with his fly-fishing partner and co-author, Jerry Hubka. Rick is a fly designer for Umpqua Feather Merchants and has published stories in *Fly Fisherman* and *Fly Tier* magazines. He is a field tester for Rio Fly Lines, Whiting Farms Pro Staff, and Daiichi Pro Staff.

"Help! I can't swim!" An angler on Alaska's American Creek prepares to drop a Morrish Mouse in the sweet spot.
BRIAN O'KEEFE

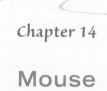

Chapter 14

Mouse

Ken Morrish

Technically speaking, mice are not a member of the insect family. Yet anyone who has fished the wild rainbow rivers of Kamchatka or Alaska—or chased big browns at night in Michigan, Pennsylvania, Wyoming, and beyond—knows that the furry protein package posed by a mouse, vole, or lemming brings trout to the surface with reckless abandon. Though when Ken Morrish began guiding in Alaska circa 1990, he had his doubts.

"I'd always known that some anglers fished with mouse patterns up in Alaska, particularly on the Kanektok," Ken began. "But at the time, I thought that fishing with mice was for freaks. I equated it with my mistaken impressions of fly fishing skating patterns for steelhead—sure, you might catch one now and again, but probably not. I never fished mice then, and I look back upon those days as a time of missed opportunities. When I started Fly Water Travel in 1999, I was fortunate to visit Kamchatka in the early days, and I became more curious about the possibilities of mouse flies. I was familiar with what was out there, and there were aspects of the flies that I didn't care for, just from looking at them. The patterns weren't engineered to stay on the surface properly and were cumbersome to cast.

"When I set out to make the first prototypes of what would become the Morrish Mouse, I had a simple maxim: maximum profile, minimal mass. I wanted a fly that could be thrown with a 5-weight if necessary, even though it would most likely be cast with a 6- or 7-weight. I also wanted a fly that would land upright and resist diving down. I felt it should have a wiggly tail. It needed to look big to fish, but cast small. I didn't think ears or whiskers were important. The original prototype was a little shaggier than today's production model. Initially, I tied them with tan foam, but then I couldn't source it, so it was switched to black. That's turned out to be better. The Morrish Mouse has turned out to be the simplest production pattern I've ever made. It's a three-step fly: bunny tail, elk hair, and foam backstrap.

"I tied the mouse in anticipation of bringing one of the first commercial groups over to Russia's Two Yurt River to fish with Will Blair, who was one of the pioneers of fly fishing in Kamchatka. It was, needless to say, a fabulous adventure. On the first day we fished leeches on the swing, and the fishing was really good. On the second day we fished a section of the river that was shallower, and the going was tough. After a few hours of slow fishing, Will suggested that we give the mouse a go. Several of us tied them on, and the fish lit up. The mouse patterns proved more effective than the wet flies, and from that point on we fished all mice, all the time, until we'd worn out the flies we had. By the end of the trip, we were down to fishing steelhead skaters and Bombers, as they were the only flies left that would throw a wake.

"That first experience in Kamchatka was extremely important for me. It showed me how well a mouse could work, how the fish reacted to it, and how game they were to engulf the big bug. It gave me great confidence, and I've been a huge mousing advocate ever since."

Help me! I can't swim much longer!

These are the words we might imagine our mouse pattern squeaking as it sputters across a riffle or pops through a slick beneath some willows. Assuming that you begin with a level of confidence that some aggressive (and presumably large!) trout will see your mouse as a morsel and not a bad joke, there are few more-exciting ways to fish—especially if you're on water that has potential to yield a double-digit rainbow. Is there a time when you should think about tying a mouse on when you're in the greater Bristol Bay region? Some say late June, before the first runs of salmon (and their accompanying eggs) have arrived. Others say late July, when water levels have dropped, water temps have risen, and the vanguard of the sockeye run have begun moving into smaller, grass-bordered creeks.

Ken thinks anytime is a good time for a little mousing. "I fish them in situations where people think that it's a horrible idea," he continued. "I believe that many of us *think* we know how things work, so we fail to experiment and learn. Personally, I might catch three fish using eggs or flesh and then say, 'I guess it's time to try the mouse.' I might be in the middle of a stretch of spawning salmon where my friends are fishing beads, and I'll skate a mouse through. More often than not, I'll catch fish. I'm happiest if I can find braided structure or smaller streambeds. I want to bring the fly from the bank either straight or slightly downstream and across. I'll usually impart an upstream mend, as real mice swim with their head upstream. I don't pop the fly, but pulse it with the rod tip so it chugs along and looks squirmy, alive. The slower the water, the more action I give the fly. The only time I don't pulse the fly is if the water is real bouncy in its own right. Then I fish a straight swing."

One sure way to *not* find success with a mouse pattern is to repeatedly cast in an attempt to place a fly in what you perceive to be the "perfect spot." As Alaska West guide Matt Hynes has pointed out, "Trout rarely to never see a mouse soar off the water like a Harrier jet. If the mouse hits the water, fish it. Period."

"The way trout come to the mouse varies a ton," Ken added. "Some will come up and suck or sip the fly down with great subtlety. Other fish will do these super-violent refusals; they'll come at the fly quickly, then refuse to take and even tail slap the fly. Many times the fish that violently refuse the fly will come back and absolutely kill it. Sometimes the fish will pull softly on the tail—one, two, three times— then slowly turn and swallow it. This can, in rare instances, happen on the same swing!" (Those who have observed river rainbows feeding on real rodents believe the two-phase "hit and then eat" fly behavior may stem from the fish's tendency to stun the mouse first and then return to eat the dead, disabled, and/or disoriented mammal.)

Mouse BRIAN O'KEEFE

However the fish takes the mouse, the angler needs patience—*great patience*—to avoid setting the hook too soon. "In Kamchatka, we'd sometimes have a contest to see who could wait the longest before setting the hook," Ken said. "We'd delay the set for five or more seconds—and sometimes the fish would still be there.

"I've seen some great takes in my mousing days," Ken continued, "but one particularly stands out. I was fishing a mouse on a beautiful piece of water on the Two Yurt River with a buddy named Tim Austin. I was watching Tim skate his mouse across the run, glancing up to take in the blue sky, when a 22-inch rainbow shot off the water in a great, symmetrical, arching leap. The fish was brilliantly outlined against the sky. It flew 3 feet through the air and landed head down on top of the mouse."

Morris Mouse BRIAN O'KEEFE

Morrish Mouse

Hook: Tiemco 5263, #4

Thread: Black 3/0

Tail: Brown rabbit strip trimmed to end tuft

Back: Black closed-cell foam trimmed into a long taper at the rear

Body: Spun dark cow elk or suitable alternative hair, trimmed

Ken Morrish is co-founder and co-owner of Fly Water Travel (www.flywatertravel.com). An accomplished writer and photographer, his work has appeared in *Patagonia, Outside, Fly Fisherman, Fly Rod & Reel, Northwest Fly Fishing,* and other publications.

Not Just for Rainbows Anymore

Mouse patterns aren't just for oversized rainbows. Nocturnal anglers from the Manistee in Michigan to Silver Creek in Idaho will scout out likely water during daylight hours and come back in the dead of night, mice in hand, in search of Mr. Brown. A twenty- or thirty-foot cast and a few pops may do it. When the take comes, you'll know it!

One of the most far-flung—and impressive—displays of mouse power comes from Mongolia. Here, taimen—the world's largest salmonid (sometimes approaching one-hundred pounds and more)—will feed on mice, as well as prairie dogs, waterfowl, and other fish feeding on the surface. Jeff Vermillion, one of the early US outfitters to lead trips to Mongolia, shared his realization concerning the efficacy of mouse patterns for taimen:

> When we finally got there, one guy caught a taimen on his first cast—about fifteen pounds. We caught a lot of fish like that using streamers. Fun, but not exactly what we'd come for, from a fishing standpoint. At one point I was fishing a cut bank. I put away the streamers and tied on a huge mouse. I was skating the fly across the surface when a huge wake came up from nowhere, and a big taimen—the biggest we'd seen—was trying to hammer the mouse. Finally, it grabbed the fly in an explosion of water. I was too quick on the draw and pulled the fly away. One or two casts later, the fish flew three or four feet out of the water with the mouse pattern— the biggest fish I'd ever had on a fly rod.

PMDs are one of North America's most important hatches, be it on spring creeks or tailwaters like those in Montana.
BRIAN O'KEEFE

Chapter 15

Pale Morning Duns

Rick Hafele

"One could easily make the case that pale morning duns make up North America's most important hatch," entomologist Rick Hafele began, "partly because they have such a broad scope of abundance. They are present in healthy trout streams all across the continent, in numbers that make for predictable emergences—whether its *Ephemerella excrucians* in the West or related species in the East. PMDs are happy to live in any number of habitats: spring creeks like Silver Creek or the Metolius River, tailwaters like the Green in Utah or the Deschutes, and freestones, too. They are one of the most dominant critters as far as fly fishers are concerned."

If blue-winged olives are a group of many different species often incorrectly lumped as one, pale morning duns are largely one species that are inadvertently identified as many. "Though PMDs are very widespread in the western US, they all belong to the same species (*Ephemerella excrucians*). Whether you're fishing in Yellowstone or Hat Creek in northern California, it's amazing how different they look when they come off," Rick continued. "In Yellowstone, the bug might be olive-colored and correspond to a size 14 fly. On Silver Creek, it might be size 18 and bright yellow. PMDs are extremely hard to pin down. I love it when fly tiers ask what color they should tie the bug. The truth is, it's going to be a different color and a different size wherever you go—sometimes even on the same stream! I have a series of photos from the Metolius. On one rock, the nymphs range from cinnamon red to black. When they emerged, the duns varied from rusty red to bright yellow. They were both floating down the river at the same time. PMDs are a very plastic species that way. I don't know why this is, but it sure causes havoc for tiers."

One thing that you can count on with PMDs is that once the hatch starts on waters in your part of the world, it's going to stick around for a while—sometimes for as long as five or six weeks. Given its duration, the PMD hatch is one most anglers are likely to catch at some point in the late spring or summer.

PMDs are not content to befuddle anglers with their many shades and sizes. Their behavior will also leave you pulling out your hair. "*Ephemerella excrucians* usually hatch in the surface film, but the duns also emerge subsurface, up to a foot below," Rick continued. "They really struggle to get through the film, and this creates a unique situation. I saw this on the Bighorn once a few years back. A PMD hatch was on, and the fish were working something just below the surface. I tried emergers and I tried nymphs, but neither worked. It turned out that they were taking the duns that were subsurface. I put on a Partridge and Yellow soft hackle, and started hammering them. It was the perfect situation for someone who loves entomology and fly fishing. I was presented with the mystery of what the bugs were doing, and was required to sit down and figure it out. The PMD emergence keeps you guessing, keeps you on your game. It forces you to pay close attention to how fish are feeding and to learn enough about the insect's life cycle to understand their behavior. It's what I love about the sport—solving that mystery."

Rick has fished the PMD hatch on many rivers across the West. When asked to name a favorite occurrence, he looked no further than his home waters, the Deschutes. "The Deschutes is very rich in insect life," Rick explained, "but it doesn't produce lots of surface activity. To catch trout consistently, you're often nymphing. The PMD hatch is one of the emergences on the Deschutes that brings the big trout up to the surface. It generally begins in early June, once the salmon fly and golden stonefly hatches are over. It's a nice time to be on the river, as it's not blistering hot yet.

"The hatch comes off between 11 and 3, depending on how sunny it is. Many times I'll head out earlier and nymph some riffles before the surface activity gets going. A Pheasant Tail or Hare's Ear in size 16 can be dynamite. While I'm nymphing, I'll keep an eye on the flats below. It looks like marginal water, not the kind of trout habitat that calls out to you. But many of these broad flats have aquatic vegetation, and the nymphs come off those plants. It's not productive water to nymph, but the fish move in when the duns are coming up to the surface, and that's where you want to be. I may sit on the bank for a while above the flats with a pair of binoculars. As soon as I see the duns, I'll rig up with a Harrop Dun or a Comparadun. If I'm getting refusals, I'll put on a soft hackle and swing a fly down and across. You're not going to see the fly on the water; instead, you're focusing on the flash of the fish or the movement of the leader. The takes can be great. You need to be careful not to clamp down on the first run, as you'll likely pop your tippet.

"The PMD spinner falls can be great, too, though many anglers will miss them, as they occur much later. The spinners will come down either in the late afternoon or the following morning. The challenge is, when they get on the water, it's next to impossible for the angler to see them. But for the fish that are looking up and seeing the bugs silhouetted, the spinners are like little beacons of light. If you see swallows in the air and little rises but no bugs, you can bet the fish are taking spinners. And a nice thing about the spinners is that they tend to be consistent in color, at least on the Deschutes; a size 16 Rusty Spinner will usually do the trick."

The PMDs' variety of colors, sizes, and behavior will sometimes compel anglers to think outside the proverbial fly box. Rick shared one such instance from the Bighorn. "It was late May or early June, and there was a great PMD hatch coming off each day around 11:30. We were coming upon big pods of fish, but they were very selective. One day in particular, the hatch was driving me and my companions—Dave

Pale Morning Dun JOHN JURACEK

Comparadun–PMD KEITH CARLSON

Hughes and Jim Schollmeyer—nuts. Together, we had 150 years of fishing experience, and we were going through patterns left and right. Every thirty drifts we might take a fish. We knew they were taking something subsurface, but none of our PMD-oriented patterns were working. Either Dave or Jim put on a yellow size 16 LaFontaine Sparkle Pupa, and the fish started taking left and right. We all switched to the pattern, and the fishing became ridiculously easy. It's not a fly that you'd expect to work, but it had characteristics— perhaps the way it trapped air bubbles—that the fish were looking for."

Comparadun–PMD

Hook: TMC 100 #16-20

Thread: Danville Yellow 6/0

Body: Pale Morning Dun Superfine

Wing: Natural Comparadun Deer Hair

Tail: Light Ginger Spade Hackles

An aquatic biologist by trade, **Rick Hafele** published *The Complete Book of Western Hatches* with co-author Dave Hughes in 1981. Since then he has co-authored or authored several other books including *Western Mayfly Hatches* and *Nymph Fishing Rivers and Streams*. His

instructional DVDs include *Fly Fishing Large Western Rivers* (a four-volume set), *Nymph Fishing Basics, Advanced Nymph Fishing,* and *Advanced Tactics for Emergers & Dries.* Visit www.laughingrivers.com.

Some feel that the increased incidence of E. albertae *on the Madison in the last few decades has to do with warmer water temperatures.* GREG THOMAS

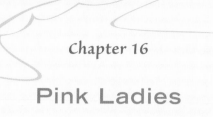

Chapter 16

Pink Ladies

Patrick Daigle

For the uninitiated, the phrase "pink lady" may summon images of a light summer cocktail, perhaps vodka, seltzer, and a splash of grenadine or Cointreau. The summer part is correct, though for Patrick Daigle the frothy water is the riffles of southwest Montana's Madison River, and the lady in question is *Epeorus albertae*.

"The pink lady emergence on the Madison is certainly not as well known as some of our other mayfly hatches, such as PMDs or Baetis," Patrick conceded. "That may be because it's usually such a trickle hatch. During the afternoon and early evening, the bugs come off very gradually. You might pick up a fish here and there with a general searching pattern, but the fish aren't very engaged. Things really get interesting when the spinner fall begins. It occurs during the last hour of daylight, and it brings up the bigger fish. If it's a strong occurrence, it's not uncommon to hook a dozen, even fifteen, trout before it's too dark to see."

Epeorus albertae are widely distributed in the West. Where its cousin *E. longmanus* prefers high-gradient, high-elevation streams, *E. albertae* prefer rivers of a more modest elevation and speed. Pink ladies, it should be noted, are not pink; instead, females are yellow with a *pinkish* tint. "Most folks lump *Epeorus* into the sulphur category," Patrick continued. "In the

East, the dominant *Epeorus* mayfly—*E. pleuralis*—is imitated with the famous Quill Gordon. The *Epeorus* wasn't much of a hatch twenty years ago on the Madison. We believe the *Epeorus* population has increased over the years due to warmer weather . . . warm winters with low snowpacks have also contributed to warmer water temperatures, allowing the insect to flourish."

Patrick might not have found his way to the *Epeorus* of the Madison had it not been for a junior high vocational ed teacher named Bill Tabacinski. "Mr. Tabacinski covered many subjects, including fisheries," Patrick recalled, "and during the fisheries segment of his natural resources course, he showed us how to tie flies. The class would have field trips to stock trout with the Connecticut DEP annually, and he would also hire a party boat, *Blackhawk,* to take all of the students out of class for a fishing trip on Long Island Sound for bluefish and stripers.

"In the summer he and his wife would head out west from Connecticut to Montana. On the first day back in the fall, he'd do a slide show in class of the two of them catching trout in Montana, with elk and bison in the background. He really opened our eyes. He and I connected, and I would tie flies at his house. I remember him taking me upriver to the Salmon River fly-fishing-only area in the town of East Hampton with my little L.L.Bean combo rod. He set me up on the best hole, and I caught a brace of fish with a Gray Ghost streamer. I was mesmerized by the beauty of the flies and the bugs. When I graduated from high school, he invited me to join him out in Montana. That was my first taste of western fly fishing, and I knew I'd found my passion. It took another seven or eight years, but I eventually made the move to southwestern Montana. Now Bill comes out some summers, and I guide him out on the Madison. He and his wife, Elaine, get the biggest kick out of giving me grief on slow days—'I thought you were a professional!'"

Patrick set the stage for a typical pink lady spinner fall on the Madison. "It's usually somewhere between mid-July and mid-August, around 9:30 in the evening. You can tell it's getting close when you see anglers starting to gather at Three Dollar Bridge! I like to get out earlier, maybe around 6, either in the Raynolds Pass area or at Three Dollar Bridge. This time of year you can anticipate that caddis might come off, so I'll usually start there. Depending on the weather, it could stay a caddis game all evening, but with a little luck, the pink lady spinner fall will start. It's a little tough on the eyes to go from a caddis pattern to a low-profile spinner. I try to make sure to switch over before it's too dark. One of Craig Matthews's Epeorus Spinners will work much of the time. The volume of spinners on any given evening can be mind-blowing; during a strong enough spinner concentration, you can hear them flying! That brings a smile to your face. As the swallows start swooping closer and closer to the water, heads start coming up. The spinners have fairly long wings and extremely long tails. They're pink, almost a Hendrickson pink. The body appears almost clear. You can pursue each head you see, and have a decent chance to catch it. As you're trying to detect rise forms, the mountains in the background are illuminated purple. It's a magical time to be on the Madison."

There are occasions, however, when the normal magic may not be quite enough to turn fortune in the angler's favor. Patrick described one such occasion, and his creative response: "I was up at Three Dollar Bridge early one evening, and there was a thunderstorm looming over Raynolds Pass. I mentioned earlier that the pink lady hatch is predominantly a spinner fall game. The one exception can be when a late afternoon thunderstorm comes through. For some reason, it concentrates the bug emergence, and the fish greedily eat the duns. This evening, the pink ladies had a beaten-up appearance as they settled on the water; they looked almost drunk, like they'd just had a long night at the Grizzly Bar. They also had a very obvious pink thorax. The bugs didn't have their typical sexy mayfly silhouette. In retrospect, I think the wind sweeping down from the pass had damaged their wings, which were still soft from their emergence. [Pink ladies leave their shucks subsurface and emerge as duns.] Whatever was going on, the Sparkle Dun wasn't fooling any fish among the disheveled, discombobulated duns on the water. I tied on a Pale Morning Dun soft hackle dressed with some floatant so it would rest in the film and found a few fish, but it wasn't quite right.

"That night I went to the vise to see if I couldn't mimic the disoriented *Epeorus* I'd witnessed earlier—a kind of buoyant soft hackle. I used some wood duck

Pink Lady JEREMY ALLAN

Daigle's Epeorus Soft Hackle Dun PATRICK DAIGLE

for the tail, tied some yellow olive for the abdomen, pink in the thorax, and ended with mallard shoulder feather to give it a beefy collar. It became the Daigle's Epeorus Soft Hackle Dun."

Daigle's Epeorus Soft Hackle Dun
Hook: Tiemco 100, #16

Thread: Rusty dun 8/0 UNI-Thread

Tail: Wood duck flank or dyed mallard flank

Body: Abdomen: PMD and Cinnamon Caddis Superfine; thorax: PMD and Cinnamon Caddis Superfine Spinner

Underwing: Medium dun Z-lon

Collar: Mallard shoulder feather (proportioned 1 to 2 times larger than normal for drunken look and buoyancy)

Patrick Daigle moved to Montana in his mid-20s, eventually landing at Blue Ribbon Flies in West Yellowstone, where he's a commercial tier. In season, he guides on the Madison, the backcountry of Yellowstone Park, and other nearby streams. His patterns include the Spent Spruce Moth, Flying Tiger Ant, Skittering Nocturnal Stone, Diving Sculpin, and GM Nymph.

Preserving the Madison

Beyond its prolific hatches, the Madison holds another tale, a story of conservation and redemption. In the early 1990s, the plague of whirling disease afflicted the Madison, and many fish were lost. Thanks in part to information campaigns targeting the angling community, the rapid spread of whirling disease has been checked, and fish populations are steadily edging toward pre-whirling disease numbers.

Like so many western rivers, the Madison has been subject to water drawdowns and shoreline degradation brought on by agricultural and cattle interests. Conservation and sporting groups have worked diligently to preserve riparian habitat by consulting with ranchers and buying up land whenever possible. The $3.00 Bridge Project is one example of how concerned anglers can take control of their destiny. In this case, a rancher needed to sell his land, which adjoined 1.5 miles of river frontage. Craig Mathews and other members of the $3.00 Bridge coalition raised the money necessary to purchase the land in question, making it forever protected and open to fishing.

"Everybody that I pass on the river thanks me," Craig recalled. "I'll always remember one occasion in particular. A six-year-old girl came up to me in Blue Ribbon Flies and gave me fifty cents. She said, 'Listen, Mr. Matthews, I stole this from my dad because he loves the river. I think I can get fifty cents more.' Even the developers who work in the area pitched in to help."

The Gunnison River's steep canyon walls create unique casting opportunities during its epic salmon fly emergence.
ANGUS DRUMMOND

Chapter 17

Salmon Flies

Jason Yeager

It's safe to say that the first salmon fly hatch Jason Yeager ever encountered on southwest Colorado's Gunnison River rocked his world. "It was like an Alfred Hitchcock movie and a biblical event all rolled into one," he recalled. "It felt like the banks were being invaded when the stoneflies began marching up to shore—a storming of Normandy in miniature. The large trout are aware of the onslaught, too. At the peak of the hatch, you begin to wonder if there's a 10- or 12-inch fish in the river. It feels ridiculous to be fishing a bug as big as the ones we'll use, and it's only made crazier by the fact that around every corner, you may encounter the biggest trout you've ever caught. There are 30-inch trout out there, though they're not easy to find. I've fished all over the world, and I have to say that this is one of the most unique and exciting fly-fishing experiences out there."

Salmon flies (*Pteronarcys californica*) are cause for excitement on the many western rivers where they emerge. Ranging from 2 to 3 inches in length and boasting an even longer wingspan, they represent serious protein; on many systems, the biggest bug of the year. When the river temperatures hit the low 50s (exact temperature may vary system to system), the big black salmon fly nymphs (well represented in the Kaufmann's Stone) begin their trek to the shore and plant themselves on rocks, tree trunks, grass, and anywhere else they can gain purchase. The bright orange creature that leaves the black shuck is a sight to behold. The adults may crawl about for several weeks in search of a mate. During this time the wind that may sometimes dog one's cast becomes a friend, blowing the hapless bugs into the water. It's then—and when the females return to the water to drop their eggs—that the top-water feeding bacchanalia begins.

"Salmon flies are very clumsy," Jason continued. "Often when they crawl out on the rocks at the side of the river, or on the cliffs that border the Gunnison, they'll tumble back in, and then flutter their wings as they swim back toward the rocks. The fish are conditioned to seeing bugs floating right by the bank or the cliff. There are feeding lanes that the fish only occupy during the salmon fly hatch to monopolize on the big bug's behavior. All the biggest fish are in those lanes; the smaller fish don't stand a chance."

Casting big flies for big fish (who temporarily forget the wariness that made them big in the heat of the hatch) is a thrill wherever salmon flies come off—on the Big Hole, the Madison, the Henry's Fork, the Deschutes. The unique setting of the gorge section of the Gunnison enhances the experience. In places, cliff walls climb nearly 1,000 feet up from the river bed, creating dramatic and idiosyncratic casting opportunities.

"Many of the prime lies are in little grottos in the cliff side or in eddies created by rocks that extend into the water," Jason explained. "There are lots of cool casting opportunities where you have to sidearm the fly in. Many of the feeding lanes are right next to the cliff, and often the best way to get a good drift is to bang your fly up against the wall. For that reason, we use really heavy tippet—15- or 20-pound test. The fish don't care, and the heavier test protects against abrasion from the rocks and allows anglers to land fish quicker . . . and gives you a chance to land one of big hogs if you hook one! As the hatch progresses, the fish sometimes turn off on eating the bugs on the water. However, they get used to seeing them on the cliff walls and grassy banks, and will sometimes stick their heads out and grab them off. It's not easy to make that presentation—4 to 6 inches off the water—but if you can cast so the end of your tippet and fly dangles off a little vegetation, it can be done.

"There was one occasion where I was fishing upstream with a client and saw a large brown off the point of a cliff. The water was coming off the point, forming a little eddy near the cliff, and the brown was holding tight to the cliff in the shade line. We came in below the fish and watched 200 or 300 naturals go by in the eddy line a few feet away, and he wouldn't look at them. The situation was compounded by the fact that fishing to this fish from below was almost impossible, because the angler was in a precarious position on the cliff, downstream and around the corner of the cliff, unable to see the fish or his fly. Upstream access was limited by the cliff, so that wasn't an option. Not exactly an occasion for hope, but we had to try. Many times these large fish will take a fly presented right along the wall, while avoiding hundreds that float by, so we figured we had a chance. We spent twenty minutes on that fish. I put the angler on his perch on the cliff while I spotted from below. After a number of casts, the client got the fly into the eddy, drifting right down the wall. I could see the fish tracking it for 4 or 5 feet, when the fish finally decided it was the real deal. The trout's whole head came out of the water, and it ate the fly right off the rock. I waited a second and then told him to strike. It was such a slow grab, I think he would've pulled the fly away if he had been able to see the take."

Jason is the first to admit that there are times during the salmon fly hatch when the fish are so uninhibited that pattern choice may have little bearing on success. "It's definitely presentation over imitation on the Gunnison. But there are a few things I look for in

Salmon Flies ANGUS DRUMMOND

Tantrum KEITH CARLSON

a pattern: silhouette, floatation, and visibility. When I first started guiding, people were using Sofa Pillows and Stimulators—ties that relied on the hackle to keep them afloat. I found that my clients were spending half their time false casting to try to dry their flies off. That's time the fly isn't in the water, fishing. I started experimenting with foam. This is back in the late '90s, when it wasn't as popular. I remember encountering some anglers that didn't think the foam flies would be effective, but many changed their minds when they gave my flies a shot. The foam flies had a realistic silhouette and would float, even in the rapids. I also wanted the fly to be really visible. If you can't see your fly, you're riding the bench, hoping to get in the game.

"My favorite fly for the salmon fly hatch on the Gunnison is called the Tantrum. The name was sort of a joke; people out on the river will often ask, 'What are you throwing?' and someone suggested, 'A tantrum!' The Tantrum is a cross between two great flies, the Chernobyl Ant and Turk's Tarantula. In my experience, these flies tend to get a little waterlogged or are difficult to see in broken water or difficult light. I tried to come up with something that floats high, is visible, durable, and offers a credible silhouette. The fly looks a lot like a fluttering stonefly with the calf tail wing and active rubber legs. In fact, last year two of the largest fish caught all year in my boat ate the Tantrum."

Tantrum

Hook: Long dry fly, #2X

Body: 2mm foam in olive, orange, or red with 1mm black foam overbody

Wing: Krystal Flash, calf tail, and Wingbrite Hi-Vis Poly Yarn

Head: Spun and clipped deer hair

Legs: Barred rubber legs

Jason Yeager started fishing on the banks of the Rio Grande as a young boy, beginning a passion that has taken him all over the globe. Today, he guides for Gunnison River Expeditions in Colorado and is a signature tier for Idylwilde Flies.

The Skwala is the first "large bug" occurrence on some western rivers, including Montana's Bitterroot. JOHN HERZER

Chapter 18

Skwala Stoneflies

Greg Thomas

If you live in a western town with proximity to a good trout stream or two, the Skwala stonefly is the kind of hatch you might see headlined in the local newspaper. (If you're incredulous, just look at some back issues of the *Missoulian*.) While some may learn of the Skwala emergence in the newspapers, Greg Thomas came upon the hatch as part of a kind of journalism project.

"It was the mid-'80s, and I'd moved from Seattle to Missoula to attend college and play basketball for the University of Montana Grizzlies," Greg began. "That first fall, I wrecked my ankle—tore all the ligaments. It was pretty apparent I wasn't going to be going pro, so I set basketball aside and told people I was going to focus on journalism. My greater focus was on Rock Creek [about 15 miles east of Missoula]. I was up there almost every day in the winter, mostly throwing little midges and Baetis. I was away so much, I almost got thrown out of school. I wasn't having great success, but Doug Persico, who used to own Rock Creek Fisherman's Mercantile [he passed away in 2010] told me to hang in there, that spring fishing was the best. I hung in there, and Doug was right. Two good hatches arrived in the early spring—the March brown and the Skwala stonefly. As I began to branch out from Rock Creek in my fishing, I found that the Skwala emergence on the

Bitterroot River was really, really good. I found success in terms of both numbers of fish and size of fish, and began spending most of my fishing time down there."

If one river is synonymous with the Skwala stonefly, it's the Bitterroot. The river flows in a northerly direction through its namesake valley, framed by the Sapphire Mountains to the east and the Bitterroots to the west along the Idaho border; it empties into the Clark Fork near Missoula. The upper stretches flow through thickly wooded countryside; in its lower reaches, the Bitterroot courses through farmland and as you near Missoula, past homesites. Where most Montana rivers begin to see their "hatch" of out-of-state anglers in late June and July, the Bitterroot sees more early-season fishermen, thanks in part to the promise of its big bugs.

There's no question that the Skwala hatch is significant in western Montana (and, as Greg has learned, in other western states). But its status among emergences may have as much to do with its timing as its fecundity. "We have pretty rough winters," Greg ventured, "and you're either a winter angler or you're not. When the Skwala arrives, it's on the heels of some better weather—daytime temperatures might be pushing into the 40s or even the 50s. You're not chipping ice off your guides, which in itself is enough to bring a smile to many a winter angler's face. The trout are happy, too; after months of eating midge larvae, they have something big to eat. When you have a chance to put away the strike indicators and throw a size 6 or 8 dry fly and attract the attention of trout ranging from 12 to 24 inches, it's something that you should treasure . . . especially when it comes after a long winter. Even when the fishing isn't great, the Skwala is an excuse to get out, a celebration of the departure of winter and the coming of spring and summer." (Greg warns that though spring may be in the air, cold winds blowing sideways snow can always be lurking around the corner in Montana—whatever month of the year it is!)

Just when the Skwala hatch will occur is Mother Nature's best guess. (Skwalas, incidentally, resemble salmon flies in many respects, though are more understated in color and smaller. They're actually more closely related to little yellow stoneflies.) Nymphs shed their exoskeletons and begin their march toward shore when water temperatures nudge 42 degrees. "The true dry-fly junkies around Missoula begin looking for the hatch at the end of February," Greg continued. "In an unseasonably warm year, you might see some bugs then. It usually gets pretty good by mid-March and can last through April—and even later. A few years back, I was fishing in May and expecting to throw streamers or maybe see some Baetis. Instead, there was a Skwala hatch like I'd never seen. The bugs were blanketing the bankside brush, and in the afternoon the females started flying and dropping eggs. It looked like it was raining."

Skwala stones are accommodating bugs on several levels. "For starters, it tends to be an early afternoon occurrence," Greg offered. "You can sleep in a bit, have that fourth cup of coffee at breakfast, and still be on time. It generally is over by late afternoon, so you'll likely be able to get home in time for dinner . . . or still make happy hour. I've found that the fish are not particularly discriminating about the patterns that they'll take. Olive Stimulators will work fine. One of the tried-and-true patterns for me is the Peacock Trude, tied by Chuck Stranahan, who guides on the Bitterroot and has a long association with the river. One mistake people make with their Skwala ties is overdressing them. The naturals ride low in the water; many patterns are too big and bushy. I like patterns that are a little more subtle. I also like to see a little sparkle in the wing. Light radiates through, and this simulates the flutter of the bug's wings when they're trying to lift off the water."

Thanks to their protein mass, stoneflies can bring out the best (or should we say, least cautious) behavior in trout. Still, one should manage expectations—at least on the Bitterroot. "The rivers of western Montana are not as fertile as those in southwest Montana, and the fish numbers reflect that," Greg explained. "When the hatch is coming off, you won't see fish rising in all stretches. The trout are going to be grouped together in runs that give them good cover and feeding opportunities, and these might be a mile apart. For that reason, floating is more productive than wade fishing. On a good day during the Skwala hatch, you can expect to get six fish to the net, with that many more misses on top."

There are times—even with a hatch of big bugs like the Skwala—when you can't find a fish no matter what you do. Greg recalled such a day: "We were on the Bitterroot, and all the conditions were right.

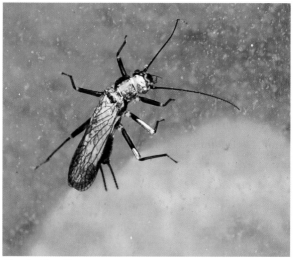

Skwala Stonefly GREG THOMAS

Skwala Stonefly BRIAN O'KEEFE

There were bugs on the water and few other anglers were about, but we couldn't catch a thing. My friend was rowing and I was casting from the front of the raft. It confused me—I knew the hatch was on, but no fish were responding . . . *at all.*

"We rounded a corner of the river, and there was a raft pulled over with a couple long poles sticking out of its bow. Two men were kneeling in the shallows. It was Montana Department of Fish, Wildlife & Parks, and they'd been electroshocking the river, doing fish counts. We pulled over to say hello. 'It was you guys!' we joked. 'No wonder we weren't catching them!' As soon as we got downstream of the Fish & Wildlife guys, the fishing turned on."

Greg Thomas is the editor of *Fly Rod & Reel*. Before taking the helm at *FR&R*, he served as the managing editor of *Big Sky Journal* and was the founder and publisher of *Tight Lines*, an annual publication he sold in 2007. Greg also operates Angler's Tonic (www.anglers tonic.com), an edgy fly-fishing website.

Missoula—The Other Montana

When out-of-state anglers plan a trip to Montana, they usually set their sights for the southwest corner of the state—the Yellowstone, the Madison, the Big Hole, the spring creeks of the Paradise and Ruby Valleys. These are classic waters, for sure. But for a change of scene—and some equally impressive angling—consider setting your sights a bit north toward Missoula. Not only will you find the Bitterroot and Rock Creek (as Greg shared), but miles and miles of the Clark's Fork, the intimate pleasures of Fish Creek, and a host of cultural attractions.

Oh yes—and the Blackfoot.

Even the most casual reader of fly-fishing literature has likely come across Norman Maclean's classic *A River Runs Through It,* and has imagined him- or herself casting next to Paul in the Big Blackfoot. If you missed the book, you might have caught the movie. (Oddly enough, the film was shot on and around the Gallatin, which looks nothing like the Blackfoot.)

"When I first started and would exhibit at an outdoor show out of state, people would come by the booth and say 'Missoula—oh, where's that?'" recalled John Herzer, co-owner of Blackfoot River Outfitters in Missoula. "All that changed with the movie. I wouldn't say that we had a bunch of newbies coming to Missoula, asking to fish the Blackfoot River. Instead we got a lot of accomplished anglers who were perhaps already coming to Montana to fish, and wanted to try our region. People realized that we had four-hundred miles of river, without the crowds."

The Upper Delaware is a prolific wild trout fishery that, amazingly, is within five hours' drive of nearly 25 percent of the population of the United States! AARON LUCAS

Chapter 19

Sulphurs

Paul Weamer

If Hendricksons are the first major hatch of the season in the East for many anglers, sulphurs are the first major *evening* hatch. "In early spring when the Hendricksons are popping, it can still be cold and snowing out," Paul Weamer ventured. "One nice thing about sulphurs is that they usually arrive here in Pennsylvania in mid-May. It's peak season, and the bugs hatch in prolific numbers . . . and the fish respond in kind. Every part of the hatch is productive—nymphs, emergers, duns, and spinner falls. It might be our most important mayfly. Even if I'm working late at the fly shop, I can take comfort that the hatch will still be coming off, and I'll have an hour or two to fish before it's dark."

When you speak of sulphurs in fishing circles, it can mean several things. In the earlier part of the season, it means big sulphurs—*Ephemerella invaria*—the most common of the sulphur hatches. "Big" is relative; these bugs are usually the size of a #14 or #16 fly, with three tails and gray wings.

"One reason big sulphurs are so common is that they are one of the most pollution-tolerant mayflies," Paul continued, "and that's significant in the East, where water quality may be more compromised. Case in point: I used to live on the edge of a small coal town in the mountains near Altoona, Pennsylvania. A cold but blemished trout stream flows near the house where I lived. Above town, pungent fluorescent-orange tributaries discharge acidic mine effluent into the stream; polluted water eventually paints the streambed a sickly carrot color approximately 10 miles below town. But I lived in the space between the orange stretches, where a cool spring day streamside could make you forget what lies above and below town. One day I stopped at the bridge on my way home from work to look at the stream. I was shocked—small, pale insects fluttered off the water and raced toward streamside vegetation, trying to avoid the dive-bombing swallows. I could tell from my perch on top of the bridge that they were mayflies, and I was so excited I climbed down a muddy bank in my good shoes to try to identify them. I remember thinking, 'If these things can live here, they can live anywhere.' They were sulphurs, hundreds of them. And trout were rising to eat them, just as if men and coal mines had never existed."

The little sulphur (*Ephemerella dorothea dorothea*) is the second most common sulphur occurrence. They begin to show up in late May and may overlap with their larger relatives; they are the equivalent of a #18 fly. On Paul's old stomping grounds on the Upper Delaware, they are a mayfly staple and can linger through August. "On some summer days, afternoon emergences will blanket the river," Paul said, "even when the temperature hits 90 degrees. I believe the afternoon hatch is a result of cold-water releases from the dams on the East and West Branches of the river. Little sulphurs are more commonly an evening occurrence on limestone and freestone streams. I've seen mating clouds of spinners so dense that it appears as though there's a fog suspended over the riffles."

Nestled among the Catskills separating New York and Pennsylvania, the Upper Delaware River system comprises over 80 miles of wild trout water, encompassing the tailwaters of the East Branch and West Branch and the first 27 miles of the main stem, to Callicoon, New York. Where many eastern trout rivers have become history, their fisheries destroyed by the effluence of an ever-infringing population, the Upper Delaware system—with the help of several 1960s-era dams—has thrived. The West Branch of the Delaware hosts healthy populations of brown trout and fishes well throughout the year; the East Branch sustains browns, rainbows, and brookies and is a great spring and fall fishery; the Big D holds rainbows that average 15 inches and deport themselves like fish half again their size. Such a fishery will turn heads anywhere; it's especially impressive when one considers that it lies within five hours' drive of nearly 25 percent of the population of the United States!

As sulphurs occur on so many different river systems, it's no wonder that their color ranges tremendously. Paul explained his approach to overcoming color challenges. "There's no question that color is very important when you're trying to match sulphurs. Many people tie bugs in a solid color, but in nature the mayflies don't appear this way. One day I walked into the shop and noticed all these bucktails in every shade imaginable. I thought it would be worth a try to take individual fibers from different shades of bucktail and mix them together. The first few I tied, I was fooling around. But I found that I really liked the way they looked; they seemed very natural in the way they suggested several body-color shades in one fly. There's a one-bug tournament on the Upper Delaware in late April, and I tied a few of these bucktail-imbued flies to imitate the Hendricksons that were coming off at this time. I won the tournament with one of those flies, and I was sold. These early efforts led to the Bucktail Parachute, a pattern that works particularly well for the big sulphurs. The flies float well when they are greased, and you can create infinite body colors by combining different strands of dyed bucktail. The twisted bucktail fibers also give the flies a realistic segmented appearance."

Paul's Bucktail Parachute Sulphurs have proven their mettle on many streams. One of the fly's most memorable conquests came on the Delaware: "The

Sulfur LUCAS CARROLL

biggest fish tend to hold in screwy areas where it's hard to get the drag right. One afternoon, I came upon two good fish: a 24-inch fish in a little eddy gulping bugs as they circulated upstream, and an 18-inch fish in a mini riffle just behind it. As the sulphur hatch progresses through the summer, the fish see every sulphur pattern ever created, and they get very picky. I tried a bunch of sulphurs, and couldn't even get a look. I thought I'd try the Bucktail Parachute. As I was casting, something spooked the smaller fish. It swam forward and bumped

Bucktail Parachute Sulfur PAUL WEAMER

into the bigger fish, and the bigger fish moved into the riffle. My fly happened to land in the riffle, and on the first drift, the 24-inch fish ate it. I owe the 18-inch fish behind it at least half of the credit."

Bucktail Parachute Sulphur
Hook: Daiichi 1170, #14 or #16
Thread: Light Cahill 8/0 UNI-Thread
Tail: Brown bucktail fibers
Abdomen: 3 yellow, 2 orange, and 1 olive bucktail fibers, twisted together
Wing: Medium dun Darlon or Antron
Hackle: Barred ginger
Thorax: Yellowish-orange beaver fur dubbing

Paul Weamer is a contract fly designer for the Montana Fly Company and the inventor of the Weamer's Truform, Comparachute, and Alewife and the Weamer Streamer series of flies. He is *Fly Fisherman* magazine's Northeast field editor. His first book was *Fly-Fishing Guide to the Upper Delaware River.* He also co-authored *Pocketguide to Pennsylvania Hatches* (with Charles Meck) and contributed to Jay Nichols's *Tying Dry Flies,* among other books.

The Cradle of American Fly Fishing

Ask the average contemporary trout angler to point you toward America's fly-fishing mecca, and nine out of ten will steer you in the direction of southwest Montana. Had you asked that question one hundred years ago, the answer would have been some two thousand miles east—along the streams in and near southern New York's Catskills Mountains. It was here in the town of Central Valley that rod maker Hiram Leonard set up shop in the early 1880s, after perfecting hexagonal bamboo rods in his home state of Maine. (His secrets? A beveling machine that he designed, which cut the six strips needed for each taper, and the use of formulas to create rods with compound tapers.)

Not long after, fly tier Theodore Gordon ushered in a new era of American fly fishing with his realistic dry-fly creations, using the Beaverkill, Neversink, Esopus, and Upper Delaware as his laboratories. At the time, most American fly anglers used wet flies. Gordon imported dry flies from his friend Frederick Halford in England and adopted them to better match the bugs he encountered in upstate New York. His most famous creation, of course, was the Quill Gordon.

Where many eastern trout rivers have become history, their fisheries destroyed by the effluence of an ever-infringing population, the Upper Delaware system—with the help of several 1960s-era dams—has thrived. That such a fishery endures is impressive—especially when one considers that it lies within a five-hour drive of nearly one-third of the population of the United States!

The beautiful Battenkill—in the backyard of Orvis's Manchester, Vermont, headquarters—reveals its secrets slowly, and can be doubly frustrating during summer Trico hatches. TOM ROSENBAUER